GROUP 2nd Edition
DYNAMICS
for TEAMS

GROUP 2nd Edition
DYNAMICS
for TEAMS

Daniel Levi
Cal Poly, San Luis Obispo

SAGE Publications
Los Angeles · London · New Delhi · Singapore

For information:

Sage Publications, Inc.
2455 Teller Road
Thousand Oaks, California 91320
E-mail: order@sagepub.com

Sage Publications Ltd.
1 Oliver's Yard
55 City Road
London EC1Y 1SP
United Kingdom

Sage Publications India Pvt. Ltd.
B-42, Panchsheel Enclave
Post Box 4109
New Delhi 110 017 India

Printed in the United States of America

Library of Congress Cataloging-in-Publication Data

Levi, Daniel, PhD
Group dynamics for teams / Daniel Levi.—2nd ed.
   p. cm.
Includes bibliographical references and index.
ISBN-13: 978-1-4129-3749-8 (pbk.)
   1. Teams in the workplace. I. Title.

HD66.L468 2007
658.4'022—dc22   2006030619

This book is printed on acid-free paper.

09   10   10 9 8 7 6 5 4

Acquisitions Editor:	Cheri Dellelo
Editorial Assistant:	Anna Mesick
Production Editor:	Libby Larson
Copy Editor:	Barbara Ray
Typesetter:	C&M Digitals (P) Ltd.
Proofreader:	Word Wise Webb
Indexer:	Jeanne Busemeyer
Cover Designer:	Glenn Vogel

Brief Contents

Contents

Acknowledgments

Many people helped to shape the development of this book. My understanding of work teams, including both manufacturing and professional teams, was fostered by the many opportunities I had to study and consult with actual teams in industry. Andrew Young and Margaret Lawn (formerly of Nortel Networks) and Don Devito (formerly of TRW) created a number of opportunities for me to work with teams in the United States and abroad. Most of my research and consulting on work teams were performed with Charles Slem, my partner at Cal Poly, San Luis Obispo. In addition, I had the opportunity to work with engineering teams at Cal Poly as part of a NASA-supported program to improve engineering education. This project was supported by Russell Cummings, Joanne Freeman, and Unny Menon of the College of Engineering at Cal Poly as well as the engineering graduate students who worked with me, especially Lawrence Rinzel and Maria Cacapit. As a teacher of group dynamics, I learned a lot by co-teaching with Fred Stultz and Robert Christenson.

The support of people at Sage Publications has been invaluable, especially from my editors, Jim Brace-Thompson and Cheri Dellelo. Judith Barnes, Kathy Johnson, and Sara Kocher labored diligently to improve my language and make the text more readable. My wife, Sara, deserves special credit for her thoughtful reviews and supportive presence throughout this process.

Introduction

There are two sources of information about teamwork. On the one hand, there is a large body of research in psychology and the social sciences called group dynamics that examines how people work in small groups. This research has been collected over the past century and has developed into a broad base of knowledge about the operation of groups. On the other hand, the use of teams in the workplace has expanded rapidly during the past two decades. Management researchers and applied social scientists have studied this development so as to provide advice to organizations about how to make teams operate more effectively. However, these two areas of research and knowledge often operate along separate paths.

The purpose of this book is to unite these two important perspectives on how people work together. It organizes research and theories of group dynamics so that this information can be applied to the ways in which teams operate in organizations. The concepts of group dynamics are presented so that they are useful for people who work in teams and enlarge their understandings of how teams operate. It is hoped that this integration will help readers to better understand the internal dynamics of teams so that they can become more effective team leaders and members.

The larger goal of this book is to make teams more successful. Teams and groups are important in our society, and learning teamwork skills is important for individual career success. The book presents many concepts related to how groups and teams operate. In addition, the chapters contain application sections with techniques, cases studies, and activities that are designed to develop teamwork skills. Teamwork is not just something one reads about and then understands; teamwork develops through guided experience and feedback. This book provides a framework for teaching about teams and improving how teams function.

Overview

The 18 chapters in this book cover a wide range of topics related to group dynamics and teamwork. These chapters are organized into the following four parts: characteristics of teams, processes of teamwork, issues teams face, and teams at work.

Part I: Characteristics of Teams

Chapters 1 and 2 provide an introduction to group dynamics and teamwork. Chapter 1 explains the differences between groups and teams. The purpose of using teams in organizations and why they are increasing in use are examined. The chapter concludes with a brief history of both the use of teams and the study of group dynamics.

Chapter 2 explores the characteristics of successful teams. It explains the basic components that are necessary to create effective teams and examines the characteristics of successful work teams. In many ways, this chapter establishes a goal for team members, whereas the rest of the book explains how to reach that goal.

Part II: Processes of Teamwork

Chapters 3 through 6 present the underlying processes of teamwork. Chapter 3 examines the processes that relate to forming teams. Team members must be socialized or incorporated into teams. Teams must establish goals and norms (operating rules) to begin work. These are the first steps in team development.

Chapter 4 presents some of the main concepts from group dynamics that explain how teams operate. Working together as groups affects the motivation of participants both positively and negatively. Team members form social relationships with one another that help to define their identities as teams. Teams divide their tasks into different roles to coordinate their work. The actions of team members can be viewed as either task oriented or social, both of which are necessary for teams to function smoothly. One of the underlying concepts that defines teamwork is cooperation. Teams are a collection of people who work cooperatively together to accomplish goals. However, teams often are disrupted by competition. Chapter 5 explains how cooperation and competition affect the dynamics of teams.

Team members interact by communicating with one another. Chapter 6 examines the communication that occurs within teams. It describes the communication process and how teams develop communication patterns and climates. The chapter also presents practical advice on how to facilitate team meetings and develop skills that help to improve team communication.

Part III: Issues Teams Face

The third part of the book contains seven chapters that focus on a variety of issues that teams face in learning to operate effectively. Chapter 7 examines conflict in teams. Although conflict often is viewed as a negative event, certain types of conflict are both healthy and necessary for teams to succeed. The chapter explains the dynamics of conflict within teams and discusses various approaches to managing conflict in teams.

Chapter 8 describes how power and social influence operate in teams. Teams and their members have different types of power and influence tactics available to them, and the use of power has wide-ranging effects on teams. In one important sense, the essence of teams at work is a shift in power. Teams exist because their organizations are willing to shift power and control to teams.

For many types of teams, their central purpose is to make decisions. Chapter 9 examines group decision-making processes. It shows under what conditions teams are better at making decisions than are individuals and the problems that groups encounter in trying to make effective decisions. The chapter ends with a presentation of decision-making techniques that are useful for teams.

Chapter 10 presents the leadership options for teams, from authoritarian control to self-management. The various approaches to understanding leadership are reviewed, with an emphasis on leadership models that are useful for understanding team leadership. The chapter examines self-managing teams in detail to illustrate this important alternative to traditional leadership approaches.

The different approaches that teams use to solve problems are examined in Chapter 11. The chapter compares how teams solve problems with how teams should solve problems. The chapter presents a variety of problem-solving techniques that can be used to help improve how teams analyze and solve problems.

Creativity, which is one aspect of groups that often is criticized, is discussed in Chapter 12. Groups can inhibit individual creativity, but some problems require groups to develop creative solutions. The chapter examines the factors that discourage creativity in groups and presents some techniques that foster group creativity.

Chapter 13 examines how diversity affects teams. In one sense, if everyone were alike, then there would be no need for teamwork. Teams benefit from the multiple perspectives that come from diversity. However, group processes need to be managed effectively for these benefits to be realized.

Part IV: Teams at Work

The final section of the book presents a set of issues that relate to the use of teams in organizations. Chapter 14 examines the relationship between teams and the cultures of their organizations. Culture defines the underlying values and practices of a team or organization. Teams develop cultures that regulate how they operate. Work teams are more likely to be successful if their organization's culture supports them. Transnational teams need to develop a hybrid culture that mediates the cultural differences among its members.

Although teams often are thought of as people interacting directly with one another, Chapter 15 examines the impacts of teams that interact through technology. Virtual teams contain members who may be dispersed around the world and use a variety of technologies to coordinate their efforts. The use of these technologies changes some of the dynamics of how teams operate and how organizations use teams.

The different types of applications of teams in the workplace are presented in Chapter 16. Teams can be created among factory and service workers, professionals, or managers. These different types of teams create different opportunities and risks for organizations. Regardless of the types of teams, there are issues that organizations need to manage to support the use of teamwork.

Chapter 17 examines team-building, that is, the various approaches for improving how teams operate. Organizations use team-building techniques to help teams get started, overcome obstacles, and improve performance. Teamwork training helps to develop people's skills so that they can work together more effectively.

Approaches to evaluating and rewarding teams are the topics of the final chapter. One of the keys to developing effective teams is creating a mechanism to provide quality feedback to teams so they can improve their own performance. Performance evaluation systems help to provide feedback, while reward programs motivate team members to act on this information.

Learning Approaches

Learning how to work in teams is not a matter of simply reading about group dynamics. Teamwork is a set of skills that must be developed through practice and feedback. In addition to presenting information about how teams operate, this book contains three other types of material that are helpful for developing teamwork skills: application sections, case studies, and activities.

Many chapters in the book contain application sections. The purpose of these sections is to provide practical advice on applying the concepts in the chapters. These sections focus on presenting techniques rather than theories and concepts. They can be applied to the teams one belongs to or can be used with a group in a class to practice the skills.

All chapters end with cases studies and teamwork activities. The cases studies, called Team Leader's Challenge, present a difficult team problem and contain discussion questions for providing advice to the team's leaders. The cases use a variety of student and work teams. By using the concepts in the chapter, the cases can be analyzed and options for the team leaders developed.

The teamwork activities are designed to examine a topic in the chapter and include a set of discussion questions that apply what has been learned to actual teams. Some of the activities are structured discussions or small-group exercises. However, most of the activities are structured observations of how teams operate. One of the most important ways in which to improve both one's teamwork skills and the operation of teams is to learn how to be a good observer of group processes. These observation activities are designed to develop these skills.

For the observation activities, there are several options that can be used. If the observers belong to functioning teams, then they can observe the teams to which they belong. For example, a teamwork class might have students working on project teams. The observation activities can be used to study and provide feedback to the project teams. Groups also can be created in class settings and given assignments. There are many books on small-group activities that can be used to create assignments for the groups. Small-group discussions of the Team Leader Challenges provide an alternative activity to observe how groups interact. A class could use several groups with an observer assigned to each group, or a single group could perform while being surrounded by many group process observers. Finally, students could be asked to find a team that they can observe as part of an ongoing class project.

Each of the activities includes objective, activity, analysis, and discussion sections. The structure of the activities makes them suitable for homework assignments or for entries in group dynamics journals. The basic structure of the written assignments would include answering the following questions: What did you observe? How did you analyze this information? How would you apply this knowledge?

By working through the applications, cases, and activities presented here, team members will gain practical skills and knowledge that can be directly applied to improve the operations of teams and the success of teamwork.

PART I

Characteristics of Teams

1

Understanding Teams

A team is a special type of group in which people work interdependently to accomplish a goal. Organizations use many different types of teams to serve a variety of purposes. The use of teams to perform work has a long history, but during the past few decades organizational teamwork has expanded rapidly due to changes in the characteristics of workers, the nature of jobs, and the structure of organizations. The scientific study of group dynamics provides useful insights about how teams operate and how they can be improved.

Learning Objectives

1. What are the characteristics of a group?

2. How is a team different from a group?

3. How are teams used by organizations?

4. How are traditional work groups different from teams and self-managing teams?

5. Why is the use of teams by organizations increasing?

6. What are the main historical trends in the use of teams?

7. How has the study of group dynamics changed over time?

1. Defining Groups and Teams

A group is more than just a collection of people; it has distinguishing characteristics (Table 1.1). A group exists for a reason or purpose and has a goal that is shared by the group members. The people in a group are connected to one another. They recognize this connection, and it binds them together so that they collectively share what happens to fellow group members. Group members interact with one another; communication often is viewed as the central process of a group. The people in a group recognize and acknowledge their membership. Formal and informal rules and roles of the group control the interactions of group members. The people in a group influence one another, and the desire to remain in the group increases the potential for mutual influence. Finally, a group satisfies members' physical and psychological needs such that individuals are motivated to continue participation.

TABLE 1.1

Characteristics of Groups

Goal Orientation:	People joining together for some purpose, or to achieve some goal
Interdependent:	People who have some type of relationship, see connections among themselves, or believe that they share a common fate
Interpersonal Interaction:	People who communicate and interact with one another
Perception of Membership:	Recognition that there is a collective to which one belongs
Structured Relations:	Roles, rules, and norms that control people's interactions
Mutual Influence:	Impact people have on one another because of their connections
Individual Motivation:	Satisfaction of personal needs through membership in the group

SOURCE: From David W. Johnson & Frank P. Johnson. *Joining Together: Group Theory and Group Skills, 6*, reprinted with permission, Pearson Education Copyright © 2003.

From a psychological perspective, two processes define a group: social identification and social representation (Hayes, 1997). Social identification refers to the recognition that a group exists separately from others. It is the creation of a belief in "us versus them." Identification is both a cognitive process (classifying the world into categories) and an emotional process (viewing one's group as better than other groups). Social representation is the shared values, ideas, and beliefs that people have about the world. Over time, belonging to a group changes the ways its members view the world. The group develops a shared worldview through member interactions.

Most definitions of teamwork classify a team as a special type of group. To some theorists, the distinction between groups and teams is fuzzy; teams are simply groups in work settings (Parks & Sanna, 1999). Other theorists focus on how the behavior of teams differs from that of typical groups. Teams have been defined as structured groups of people working on defined common goals that require coordinated interactions to accomplish certain tasks (Forsyth, 1999). This definition emphasizes one key feature of a team—that members work together on a common project for which they all are accountable. However, other qualifiers can be used to distinguish between groups and teams.

One common distinction relates to application. Teams typically are engaged in sports or work activities. They have applied functions, and the roles of team members are related to their functions. For example, in sports teams, members have specific assigned roles, such as pitcher and shortstop. Teams usually are parts of larger organizations, and their members have specialized knowledge, skills, and abilities related to their tasks. This is why we typically do not talk about a family as a team; in a family, roles are inherited and not directly related to tasks. This distinction appears in research on groups and teams. Research on groups typically is done in laboratory settings, whereas research on teams typically is done in field studies that focus on the use of teams in the workplace (Kerr & Tindale, 2004).

"Group" is a more inclusive term than "team." Groups range in size from two to thousands, whereas teams have a narrower range of sizes. A dating couple may be considered a group but not a team. Political parties and social organizations are groups but not teams. A team typically is composed of 4 to 20 people who interact with one another directly (although this interaction may occur through computers and other communication devices). A team is not simply people who belong to the same group or who are co-acting in the same place.

Katzenbach and Smith (1993) focus on performance in their definition of teamwork. In addition to team members having a common purpose, performance goals are connected to this purpose, for which everyone in the team is held mutually accountable. They also believe that the concept of a team should be limited to a fairly small number of people with complementary skills who interact directly. This helps to distinguish teams from work groups, whose members jointly do the same tasks but do not require integration and coordination to perform the tasks.

Hayes (1997) focuses on power in her definition of teams. A team must actively cooperate to achieve its goals. For this to occur, a team must have independence, responsibility, and the power to operate. A team is not a group of people who perform a task under the rigid control of an authority figure. For a group to become a team, it must be empowered and have some authority to act on its own. In addition, team members are more likely to work together cooperatively and provide assistance to one another than are members of other types of work groups.

Because there is no firm dividing line between a group and a team, the use of these terms in this book is somewhat arbitrary. When referring to research on group dynamics, especially laboratory research, the term "group" is used. When talking about applications in work environments where people are interdependent, the term "team" is used. For the in-between cases, "group" and "team" are used interchangeably.

2. Types and Purposes of Teams

Organizations use teams in a variety of ways. Because of this variety, there are many ways of classifying teams. These classifications help to explain the psychological and organizational differences among different types of teams. One important distinction is the relationship of the team to the organization. Teams vary depending on how much power and authority they are given by organizations.

How Teams Are Used by Organizations

Teams are used to serve a variety of functions for organizations. The day-to-day operations of organizations can be shifted to teamwork (e.g., factory production teams, airline crews). Teams can be formed to provide advice and deal with special problems, such as teams created to suggest improvements in work processes. Teams help to manage coordination problems by linking different parts of organizations, such as budget or planning committees composed of members from several departments. Finally, teams can be used to change organizations by planning for the future or managing transitions.

Obviously, teams may come in mixed packages. Concurrent engineering teams are teams composed of members of an organization whose task is to oversee the design, manufacturing, and marketing of new products. Being in a concurrent engineering team is part of the day-to-day activity of people working in research and development. However, other members of the team are there on a part-time, temporary basis to deal with coordination, special problems, and implementation of change. Research and development staff may define the characteristics of a new product, while representatives from other departments may comment on issues related to production and marketing.

Sundstrom (1999a) identifies six types of work teams on the basis of the functions they perform: (1) production teams, such as factory teams, repeatedly manufacture or assemble products; (2) service teams, such as maintenance crews and food services, conduct repeated transactions with customers; (3) management teams, composed of managers who work together, plan, develop policy, or coordinate the activities of an organization; (4) project teams, such as research and engineering teams, bring experts together to perform a specific task within a defined period of time; (5) action or performing teams, such as sports teams, entertainment groups, and surgery teams, engage in brief performances that are repeated under new conditions and that require specialized skills and extensive training or preparation; and (6) parallel teams, which are temporary teams that operate outside normal work, such as employee involvement groups and advisory committees that provide suggestions or recommendations for changing an organization.

Classifying Teams

Teams can be classified in other ways than by the types of activities they perform, but there is no agreed-upon classification system for work teams (Devine, Clayton, Philips, Dunford, & Melner, 1999). Researchers have suggested classifying teams by whether they are permanent or temporary, how much internal specialization and interdependence they require, and how much integration and coordination with other parts of the organization are needed (Mohrman, 1993; Sundstrom, DeMeuse, & Futrell, 1990). One of the most important distinctions among types of teams is how much power they are allocated (Hayes, 1997). When an organization uses teams rather than individual workers to perform tasks, it is giving the teams some power and authority to control the operations of its members. This shifting of power affects leadership, decision making, and how team members' work activities are linked.

There are three options for organizing people into work groups: a traditional work group, a traditional team, or a self-managing team. The differences among these types are presented in Table 1.2. Traditional work groups are part of the organization's hierarchical system. Supervisors or managers who control the decision-making process lead these work groups. Group members typically work on independent tasks that are linked by the supervisors or work system.

TABLE 1.2

Organization of Teams

	Traditional Work Group	Traditional Team	Self-Managing Team
Power	Part of organization's hierarchy, management-controlled	Linked to organization's hierarchy, some shift of power to team	Linked to organization's hierarchy, increased power and independence
Leadership	Manager or supervisor controls	Leader has limited managerial power, selected by organization	Leader is team facilitator, selected by the team
Decision making	Authoritarian or consultative	Consultative, democratic, or consensual	Democratic or consensual
Activities or Tasks	Independent	Interdependent, coordinated by leader	Interdependent, coordinated by team members

SOURCE: Adapted from McGrath (1984).

Traditional teams are given some power and authority, so they are somewhat independent of the organization's hierarchy. Their leaders are selected by management and given some managerial power. Team leaders can use a variety of techniques for making decisions, such as using the teams to provide advice about decisions (consultative) and having the teams vote to make decisions. Team members' work activities are interdependent and coordinated by the leaders.

Self-managing teams are given significantly more power and authority than traditional work groups, and they are more independent of the organization's hierarchy. Team members typically select their leaders, so the leaders have limited power and must facilitate, rather than control, their teams' operations. The leaders must rely on democratic or consensual decision making because they have no authority to make teams accept decisions. The work of team members is highly interdependent, and all team members work together to coordinate activities.

3. Why Organizations Are Using Teams

The traditional approach to organizing people to perform a task is called Scientific Management (Taylor, 1923). In this approach, managers or technical experts analyze a task and divide it into small activity units that are performed by individuals. The system is designed such that each activity is linked, and individuals work separately to complete the entire task. It is the role of management to design the system and control the operations of the workers. It is the role of the workers to perform the specific task. In other words, managers think and control, and workers act.

This traditional approach works very well under certain conditions. It requires that the task remain the same for some time because it is difficult to change the system. It requires that the process be not too complex or easily disrupted because the workers doing routine activities are unaware of what happens in other parts of the system. It focuses on productivity and often ignores concerns about quality and customer service because these factors require more commitment to the job. It assumes that there are workers who are willing to perform routine activities under controlled situations because that is the nature of work.

Under these conditions, Scientific Management is the best approach, and the time and expense of developing teams is not needed. Teams are important when the goal is to improve the way a product is made or a service is provided, when the job is complex, when customer service and quality are important, or when rapid change is necessary. These are the conditions that create the need for teams. Because team members have more autonomy and are developing new skills, teams encourage commitment from people who want more from work than just money. Modern organizations are shifting to teamwork because of changes in the characteristics of people, jobs, and organizations.

Characteristics of Workers

Scientific Management operates under a negative set of assumptions about workers, called Theory X (McGregor, 1960). Theory X managers assume that people are basically lazy, do not like to work, want to avoid responsibility, and need coercion to be motivated. Given these assumptions, a command-and-control organizational system makes sense. These assumptions are not valid, however, for the most desirable workers.

An alternative set of managerial assumptions, called Theory Y, is based on the belief that work is a natural activity for people, that people want responsibility, and that there is a variety of ways to motivate people. Theory Y focuses on gaining commitment to the task and getting people to accept responsibility for their work. The goal of this approach is to design a job that people will want to perform, rather than trying to force people to perform a job they dislike.

The shift to a commitment-based organization based on employee responsibility, autonomy, and empowerment is one of the core notions of teamwork. This transition helps to improve the quality of people's jobs, increase internal motivation, and improve job satisfaction (Orpen, 1979). For many people, autonomy and responsibility are the most important considerations in evaluating a job (Finegan, 1993). Often, the best way to create or enhance job satisfaction in an organization is to shift to teamwork.

Job Characteristics

The changing nature of people's jobs is encouraging the use of teamwork. Many jobs are changing from routine to nonroutine work (Mohrman, Cohen, & Mohrman, 1995). Nonroutine jobs involve more complexity, interdependence, uncertainty, variety, and change than do routine jobs. Jobs of this type are difficult to manage in traditional work systems, but are well suited for teamwork.

Nonroutine jobs are found in a number of contemporary work settings. Teams are a good way to handle factory jobs that have become increasingly complex due to technology or other factors (Manufacturing Studies Board, 1986). In modern computer-oriented factories, the typical worker operating a single machine all day is being replaced by a team of workers who monitor, troubleshoot, maintain, and manage a complex and integrated work system. Because the technology is integrated, the workers must be as well.

These changes also affect professional work. Imagine designing a new product for the marketplace. Design, manufacturing, marketing, and sales of the product require expertise from a variety of disciplines and support from many parts of an organization. Given that few individuals possess all the necessary knowledge and expertise to bring a product to completion, a diversity of knowledge is gained by using a team approach. In addition, using team members from several departments enhances support within the organization for the new product. The team members help coordinate the project throughout the organization.

The complexity of a problem or task often requires multiple forms of expertise; no one person may have all the skills or knowledge to complete a task or solve a problem, but a team offers sufficient expertise to deal with the task. Complexity also implies problems that are confusing or difficult to understand and solve. Here, the value of teamwork is not in multiple forms of expertise but rather in multiple perspectives. People learn from the group interactions in teams, and this helps them to gain new perspectives in analyzing problems and developing solutions.

As jobs become more interdependent, it becomes increasingly difficult for managers to control the flow of information. Everyone needs to be aware of changes that affect their jobs and to coordinate with others in dealing with these changes. Teams become necessary to promote coordination in a rapidly changing organizational environment.

Work is becoming more varied. Increasingly, complex technical systems do not require routine operation, but do require monitoring and trouble-shooting. Changes in technology and markets require flexibility to meet new demands. Teams provide a mechanism for creating jobs that are more responsive to the changing work environment.

Organizational Issues

The rate of change in technology and other aspects of business is con-tinually increasing. Markets are expanding, and competition is global. It is difficult to keep up with these changes using traditional approaches to orga-nizational design; the changing business environment is forcing organiza-tions to change the way they operate. Communications technology allows organizations to create new ways to integrate their operations. Businesses know they need to reduce costs, improve quality, reduce the time spent cre-ating new products, improve customer service, and increase their adaptabil-ity to an increasingly competitive environment.

As organizations change to meet contemporary demands, new organizational characteristics increase the importance of teamwork (Mohrman et al., 1995). One significant new characteristic is a shift to simpler organizational hierarchies, a transition being driven by the desire to save costs and increase flexibility by reducing layers of management. To a certain extent, teams have replaced man-agers; teams now often carry out traditional management functions.

Teams provide a way to integrate and coordinate the various parts of an organization. They can do this in a more timely and cost-effective manner than traditional organizational hierarchies. Teams execute tasks better, learn faster, and change more easily than traditional work structures—all charac-teristics required by contemporary organizations.

Although teamwork in organizations has expanded dramatically, teams have not been universally successful. Teamwork has become a management fad with its own set of problems. Organizations sometimes introduce teams in situations where they may be inappropriate. Managers may implement teams without changing the organizational contexts or supplying sufficient resources or training. Organizations sometimes call groups of employees "teams" without really changing the nature of work or the organizational reward systems.

4. History of Teams and Group Dynamics

The use of teams in organizations has changed significantly over the past century. During this period, the scientific study of group dynamics became an interdisciplinary research field.

Foundations of Teamwork

Historically, there have been two major ways of organizing people for work. One approach uses a structured hierarchy and is based on the model established by the military. Command and control is the dominant theme; everyone has a single job and a single boss, and everyone's primary activity is to do what he or she is told. The alternative is a small group or family approach, which is the model for traditional farming and the manufacturing guild system. Here, the organization is fairly small, commitment is often for life, people work their way up through the system as they learn new skills, and work is a collective activity.

The Industrial Revolution of the early 1900s shifted most work organizations to the hierarchical approach and used Scientific Management to design organizations and jobs (Taylor, 1923). Jobs were simplified, so the advantages to skilled workers created by the guild system were minimized. Professionals, from accountants to engineers, were brought into the hierarchy to make sure the production system operated efficiently. It was a system that worked well, but created problems. It alienated workers, who then became increasingly difficult to motivate; it became more difficult to set up as technical systems increased in complexity; it was inflexible and difficult to change; and it was difficult to successfully incorporate new goals other than efficiency, such as quality.

The Scientific Management model of organizations began to be questioned during the 1920s and 1930s. The rise of unions and other worker organizations demonstrated that there were problems with people's relationship to work. This led to an increased interest in the social aspects of work and the development of the Human Relations movement.

The Hawthorne studies—research projects designed to examine how environmental factors such as lighting and work breaks affected work performance—inadvertently raised questions about whether social relations in work could be ignored (Mayo, 1933). The studies revealed that social factors had an important impact on performance. In some cases, because people were being studied, they tried to perform better (what social scientists now call the Hawthorne effect). In other cases, group norms limited or controlled performance. For example, studies of the "bank wiring room" showed that informal group norms had a major impact on the performance of work groups (Sundstrom, McIntyre, Halfhill, & Richards, 2000). The "group in front" frequently engaged in conversation and play but had high levels of performance, while the "group in back" engaged in play but had low levels of performance. The work groups enforced group production norms; members who worked too fast were hit on the arm by coworkers, a practice known as "binging." In addition to the substantial impact on productivity of these informal work group norms, work groups were able to effectively enforce norms, with positive or negative benefits to the organization.

Following World War II, researchers in the United States and Europe who studied the standard approach to work recognized that although the military looked like a hierarchical system, the troops actually operated using teamwork. Research showed that organizing people in teams was one way to improve operations of organizations and improve productivity.

During the 1960s and 1970s, organizational psychologists and industrial engineers refined the use of teams at work. Sociotechnical Systems Theory (STS) provided a way to analyze what people do at work and the best way of organizing them (Appelbaum & Batt, 1994). According to STS, teams should be used when jobs are technically uncertain rather than routine, when jobs are interdependent and require coordination to perform, and when the environment is turbulent and requires flexibility. Many jobs today meet these criteria.

The most famous applied example of STS was at the Volvo car facilities in Sweden. The assembly line approach to work was redesigned to be performed by semi-autonomous groups. During the 1970s, this approach became part of the Quality of Worklife movement in the United States. Although there were several successful demonstrations of the value of using teams at work in Sweden and the United States, this teamwork approach did not become popular.

The contemporary emphasis on teamwork has its origins in another change that occurred during the 1970s. The rise of Japan as a manufacturing power resulted in the distribution of high-quality inexpensive products in the global marketplace. This caused companies in the industrialized world to change their operating methods to reduce costs while increasing quality. When business experts visited Japan to see how Japanese goals had been achieved, they found that teamwork in the form of Quality Circles seemed to be the answer. Quality Circles are parallel teams of production workers and supervisors who meet to analyze problems and develop solutions to quality problems in the manufacturing process.

Throughout the 1980s, companies in the United States and Europe experimented with Quality Circle teamwork (and later Total Quality Management). The jobs performed by workers were still primarily individual, but workers were organized in teams as a way to improve quality and other aspects of production. These early efforts were primarily attempts to copy the Japanese approach. They met with mixed success, in part because of cultural differences. By the late 1980s, new approaches had been developed, and the concept of teamwork was spreading in organizations.

The quality movement launched the current emphasis on teams, but other factors have sustained it. The increased use of information technology, the downsizing of layers of management, business process re-engineering, and globalization have all contributed to the use of teams. Teamwork in U.S. companies expanded rapidly during the 1990s and included more professional and managerial teams. Current studies suggest that 85% of companies with 100 or more employees use some type of work teams (Cohen & Bailey, 1997). In addition, some organizations are restructuring and using teams as a central element in the integration of various parts of their organizations (Mohrman et al., 1995).

Foundations of Group Dynamics

An unfortunate gap exists between our understanding of work teams and the study of group dynamics. The scientific study of groups began at the turn of the 20th century with the work of Norman Triplett (Triplett, 1898). Triplett's research showed the effects of working alone versus working in a group. For example, he observed that bicycle racers who pedaled around a racetrack alone or in groups were faster in groups. This effect has been called "social facilitation" because the presence of other people facilitated (or increased) performance. (Later research showed that performance increased for well-learned skills but declined for less well-developed skills.)

Early studies in psychology had a similar perspective in that they were designed to show how groups affected individual performance or attitudes. Although this was group research, the focus was on individuals. Psychologists did not treat groups as an entity appropriate for scientific study. This perspective changed during the 1940s, however, because of the work of Kurt Lewin and his followers (Lewin, 1951). Lewin created the term "group dynamics" to show his interest in the group as a unit of study. For the first time, psychologists took the study of groups seriously, rather than simply looking at the effects of groups on individuals. The initial work on how groups operate created a new research paradigm in psychology and the social sciences. Lewin's innovations in research methods, applications, and focus still define much of the study of group dynamics.

Lewin developed a new approach to research in psychology. He began with a belief that "There is nothing so practical as a good theory" (Lewin, 1951, p. 169). His innovation was in refining how theories in psychology should be used. He developed an approach called "action research," in which scientists develop theories about how groups operate, and then use their theories in practical applications to improve the operations of groups. The process of applying the theory and evaluating its effects would refine the theory and improve the operations of groups.

One of Lewin's primary concerns was social change. He believed that it is easier to change a group than to change an individual. If the behavior of individuals is changed and the individuals return to their everyday life, the influence of the people around them will tend to reverse the behavior change. If the behavior of a group of people is changed, the group will continue to reinforce or stabilize the behavioral change. Lewin developed models of organizational change and group dynamics techniques that are still used today.

During the 1950s and 1960s, the study of group dynamics grew beyond the field of psychology to become more interdisciplinary. Researchers from sociology, anthropology, political science, speech and communication, business, and education now study aspects of groups. Although psychological research is dominated by laboratory research on how groups operate, many other fields of study emphasize applied research. Currently, the study of group dynamics is an accepted academic discipline in a number of fields. As a theoretical area of study, it is fairly stable rather than growing. However, it is growing as an applied field as more organizations become interested in using groups and teams.

Summary

Groups are more than just collections of people. Groups have goals, interdependent relationships, interactions, structured relations, and mutual influence. Individuals are aware of their membership in groups and participate in order to satisfy personal needs. Although the distinction between groups and teams is not completely clear, the term "teamwork" typically is used to describe groups that are parts of work organizations. Team members work interdependently to accomplish goals and have the power to control at least part of their operations.

Organizations are shifting away from individual work performed in hierarchical work structures and toward team-based operations. Changing goals in organizations, which must deal with the evolving work environment, are driving this change. People are demanding meaningful work, jobs are becoming increasingly complex and interdependent, and organizations are finding that they must be more flexible. All these changes encourage the use of teamwork.

Organizations use teams in a number of ways. Teams provide advice, make things or provide services, create projects, and perform specialized activities. Teams also vary according to the power they are given, their types of leadership and decision-making processes, and the tasks they perform. These factors define the differences among work groups, teams, and self-managing teams.

Working in small groups was common before the Industrial Revolution, but Scientific Management simplified jobs and created hierarchical work systems. The Hawthorne studies of the 1930s demonstrated the importance of understanding the aspects of work related to social relations. Following World War II, researchers began to experiment with work teams. Sociotechnical Systems Theory during the 1960s presented a way to analyze work and identify the need for teams. However, it was the rise of Japanese manufacturing teams during the 1980s that led to the increased use of teamwork in the United States. Paralleling this growth in the use of teams, the social sciences developed the field of group dynamics, which focuses on understanding how groups operate. Today, group dynamics is a scientific field that provides information useful in improving the operations of teams.

TEAM LEADER'S CHALLENGE—1

You have just become the manager of an insurance office with five professional agents and several clerical assistants. The office is part of a larger company located in another city. Your office handles both sales and the processing of insurance claims. The office has been traditionally organized, with the manager running the office and supervising the employees individually.

You've heard a lot about the advantages of shifting to teamwork—it's popular in the business press. Shifting to teamwork is supposed to improve customer service, make the office more responsive to changes, and improve morale. However, you've also heard that it can be difficult to create and manage teams. You are comfortable and capable as a traditional manager, but maybe you should try something new, like teamwork.

Should you try to reorganize the office into a team?

Should the team include both the professionals and clerical assistants?

How much authority or control should you maintain over the team?

ACTIVITY: GROUPS VERSUS TEAMS

Objective. There is no clear distinction between groups and teams. The purpose of this activity is to examine the implicit definitions that people have of these terms.

Activity. Create a list of groups and teams. Using Activity Worksheet 1.1, classify these examples as groups, teams, or somewhere in between groups and teams. Compare your classifications with those of other members in your group. Try to reach agreement on the classifications by discussing how you decided.

ACTIVITY WORKSHEET 1.1

Groups Versus Teams

Groups	In-Between Groups and Teams	Teams

Analysis. When your group has reached agreement, analyze the lists and develop rules to define when a group becomes a team:

1. _____

2. _____

3. _____

4. _____

Discussion. Imagine that you are working in an office and your manager decides to organize the employees into a team. Using the rules you have developed to define a team, what advice would you give to the manager about how to create a team?

2

Defining Team Success

A successful team completes its task, maintains good social relations, and promotes its members' personal and professional development. All three of these factors are important for defining success. To perform effectively, a team requires the right types of people, a task that is suitable for teamwork, good internal group processes, and a supportive organizational context. Group members need both an appropriate set of task skills and the interpersonal skills to work as a team. Although teams can perform a wide variety of tasks, appropriate team tasks require that members' work be integrated into the final products. The group process should maintain good social relations while organizing members to perform the task. Finally, the organizational context needs to support the team by promoting cooperation, providing resources, and rewarding success.

Researchers have conducted a number of studies on work teams to determine the characteristics that predict success. Successful teams have clear goals, good leadership, organizational support, appropriate task characteristics, and mutual accountability with rewards. However, the characteristics that predict team success vary depending on the type of team being studied.

Learning Objectives

1. Understand the three criteria used to define team success.

2. Why is team success more than just completing the task?

3. What factors determine whether a group has the right set of people?

4. What types of tasks are better suited for groups than individuals? Why?

5. What are the important parts of the group process?

6. How does an organization provide a supportive context for teams?

7. Understand the characteristics of successful teams.

1. Nature of Team Success

One of the prerequisites to studying and understanding teamwork is defining the nature of team success. Research on groups and teams has used a variety of measures to study the functioning of groups. Often, research examines these internal measures of group functioning and tries to relate them to external measures of team success.

Measuring the success of teamwork can be difficult. The characteristics that team members and leaders believe are important for success might not be the same characteristics that managers believe are important (Levi & Slem, 1996). Team members focus on the internal operations of the team; they look at the contributions that each member brings to the team and how well members work together. Managers focus on the team's impact on the organization; they are concerned with results, not with how the team operates. There is a danger in using too simplistic a view of success because it may focus on the wrong factors when trying to improve a team.

According to Hackman (1987), there are three primary definitions of team success, and these relate to the task, social relations, and the individual. A successful team completes its task or reaches its goals. While completing the task, team members develop social relations that help them work together and maintain the group. Participation in teamwork is personally rewarding because of the social support, the learning of new skills, or the rewards given by the organization for participation.

Completing the Task

From a management perspective, the obvious definition of team success is performance on a task. A successful team performs the task better than other ways of organizing for the task. Although this definition may seem simple, measuring the performance of teams can be difficult. For certain complex tasks, there may be no alternatives to teamwork, making it impossible to compare group and individual outcomes. For professional tasks requiring creativity or value judgments, there may be no clear ways to determine which solutions are best (Orsburn, Moran, Musselwhite, Zenger, & Perrin, 1990).

One approach to such measurement problems is to determine whether the products or outputs of the team are acceptable to the owners, customers, and team members. However, these three perspectives may not agree with each other (Spreitzer, Cohen, & Ledford, 1999).

Completing a task successfully as a team is a measure of success, but project success is not a demonstration of team success. Could the task have been completed without a team? What was the benefit of using a team for performing the task? For a particular task, there is often little advantage to using a team. In fact, there are disadvantages because time is "wasted" in developing the team rather than focusing on the task. The advantages of using a team to perform a task occur when unforeseen problems arise and when the team works together on future tasks.

If a project runs smoothly, people working individually under supervision often can perform the necessary task. If a project encounters difficulties, however, the value of a team is demonstrated by the ability of team members to use multiple perspectives to solve problems and motivate one another during the difficult period. Although a team takes time to develop, as people learn to work together they are better able to handle future projects. Many of the benefits of creating a team occur over the long run rather than during the first project the team performs.

Maintaining Social Relations

Measuring the results of a team's task performance does not completely capture the definition of team success. A successful team performs its task and then is better able to perform the next assigned task. This is the social relations, group maintenance, or viability aspect of teamwork (Sundstrom, DeMeuse, & Futrell, 1990). An important value of teamwork is building the skills and capabilities of the team and organization. For this to happen, the team must have good internal social relations, and performing in the team should encourage participants to want to work as a team in the future.

A team must develop the social relations among its members. The social interactions necessary for teamwork require group cohesion and good communication. Cohesion comes from the emotional ties that team members have with one another. Good communication depends on understanding and trust. When team members do not develop good social relations, they do not communicate well, have interpersonal problems that interfere with task performance, and are unable to reward and motivate one another. This limits the ability of the team to continue to operate.

A good example of the problem of too much focus on task performance and too little on social relations is in the computer development team described by Kidder (1981). The team successfully developed a new computer system. However, in the stress of competition and time pressure, the team members burned themselves out. At the end of the project, everyone was happy about the success, but the team members no longer wanted to work together. Was the team a success? Yes, it completed its task, but it failed to develop social relations that encouraged successful teamwork in the future. The capabilities of the team were lost at the end of the project because of its exclusive focus on the task. The organization benefited by getting a new computer system, but it did not improve its ability to use teams to successfully design future computer systems.

Individual Benefit

The third aspect of team success concerns the individual. Participating in a team should be good for the individual. Teamwork should help to improve an individual's social or interpersonal skills (Katzenbach & Smith, 1993). In the workplace, being in a team with members with different expertise or skills should broaden an employee's knowledge and make him or her more aware of other perspectives. In addition to personal development, participating in a team should further an employee's career. Successful contributions to a work team should be reflected in the employee's performance evaluations (O'Dell, 1989).

A variety of personal benefits from working in teams helps satisfy social and growth needs. People should enjoy working in teams because it increases the social and emotional support they receive. Teams can be great learning experiences. Team members share their knowledge and expertise, and as they learn how to be good team members, they also develop communication and organizational skills.

Obviously, these personal benefits are more important to some people than to others. People vary in their social needs, and those low in social needs will be less rewarded by teamwork. Some people already have good teamwork skills, whereas others are not interested in learning these skills. Also, the social and learning benefits from teamwork primarily come from successful teams. Working on dysfunctional teams may only teach members to avoid working on teams in the future.

In addition to personal benefits, participating in a team should help a worker's career in the organization. Unfortunately, this often is not the case. Most organizations focus on managing individuals rather than managing teams. Even when most of a worker's time is spent collaborating in a team, the typical performance evaluation system focuses on what an individual produces rather than on the success of the team. Being a good team player may go unrecognized, while people who distinguish themselves and stand out get rewarded. This conflict between individual and team success is a major unresolved problem for teamwork in many organizations.

2. Conditions for Team Success

The success of a team depends on four conditions (Figure 2.1). First, the team must have the right group of people to perform the task. Second, the task must be suitable for teamwork. Third, the team must combine its resources effectively to complete the task. Fourth, the organization must provide a supportive context for the team. A team can improve the way an organization operates, but teams are not always successful (Guzzo & Dickson, 1996). To achieve success, a team requires each of these four conditions.

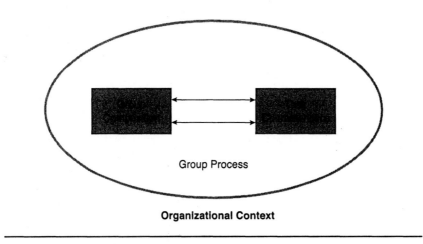

Group Process

Organizational Context

Figure 2.1   Model of Team Interaction

Group Composition

A group's performance depends on the qualities of the individuals performing the task. These qualities can be viewed in three different ways. First, the group must contain people with knowledge, skills, and abilities that match the task requirements. Second, the group must have members with the authority to represent the relevant parts of the organization and the power to implement the group's decisions. Third, the group's members must have the necessary group process skills to operate effectively.

Some groups fail because their members do not have the needed knowledge, skills, and abilities to perform their tasks. Good teams have good team members. In their study of highly successful groups, Bennis and Biederman (1997) determined that much of the success of these groups was due to the leaders' ability to recruit highly competent team members. High-performing leaders are not afraid to hire people better than themselves.

Part of creating an effective group is making sure it has the necessary diversity of knowledge and skills. Interdisciplinary research teams are more productive than teams whose members have similar backgrounds. Groups whose members have differences of opinion are more creative than like-minded groups. Management teams whose members have different backgrounds are more innovative than homogeneous teams (Guzzo & Dickson, 1996). However, diversity alone is not always a benefit to groups. The advantages of diversity are seen when members are both highly skilled and committed to their team's goals.

For some types of groups, composition is not about skills, but about representation. Task forces and other types of teams require that team members be simultaneously committed to their teams and able to represent outside interests in their teams (Gersick & Davis-Sacks, 1990). The value of such teams is their breadth of perspective and commitment to implementing decisions. For example, concurrent engineering teams include members of research, manufacturing, and other departments in the design of new products. The manufacturing representatives help ensure that the new design is sensitive to the needs of manufacturing, and their participation encourages support of the new product in manufacturing.

Concurrent engineering teams were developed to deal with the problems created by bureaucratic divisions within an organization. Before the use of this approach to design and manufacturing, the work of research and development was separate from that of manufacturing. Research and development often used the most advanced technology available because designers like to be innovative. It was not the designers' job to be concerned with production issues. The results sometimes were designs with overly expensive parts that were impossible to manufacture with the organization's existing manufacturing technologies. Concurrent engineering teams deal with these issues early in the design process.

Groups also require that team members have the skills to work together as a group. Interpersonal skills, problem-solving skills, and teamwork skills may be used as selection criteria for group members, may be taught to group members, or may be inducted through the use of facilitators (Carnevale, Gainer, & Meltzer, 1990). Interpersonal skills are communication techniques such as interviewing, active listening, providing feedback, and negotiating. Problem-solving skills improve the effectiveness of teams by providing approaches to analyzing problems and making decisions. Teamwork skills promote an understanding of group processes and provide skills to manage the group processes effectively.

Characteristics of the Task

A team's ability to perform one type of task well does not necessarily generalize to other types of tasks. The mix of abilities in the group might make it better at certain types of tasks. Teams can be used to perform a variety of types of tasks, and tasks vary in how well suited they are for group work. McGrath (1984) developed a system to describe the different types of tasks teams perform, whereas Steiner (1972) created a system that explains the different ways team members' efforts can be combined.

McGrath's system is based on the four group goals—generate, choose, negotiate, and execute—with each goal having two related tasks (Table 2.1). These eight tasks vary along two dimensions: cognitive-behavioral and cooperation-conflict. Group tasks range from purely cognitive activities (e.g., decision making) to behavioral activities (e.g., making something). Tasks also range from cooperative activities, where members work together

TABLE 2.1

McGrath's Model of Team Tasks

Goal	Tasks	Cognitive to Behavioral Dimension	Cooperation to Conflict Dimension
Generate	Planning Creative	Cognitive & Behavioral	Cooperation
Choose	Intellective Decision making	Cognitive	Cooperation & Conflict
Negotiate	Cognitive conflict Mixed motive	Cognitive & Behavioral	Conflict
Execute	Competitive Performance	Behavioral	Cooperation & Conflict

SOURCE: Adapted from McGrath, J. (1984). *Groups: Interaction and performance.* Englewood Cliffs, NJ: Prentice Hall.

(e.g., generating creative ideas), to conflict-oriented activities, where members may disagree or compete (e.g., negotiating differences).

Generation includes tasks that focus on the creative generation of new ideas and tasks that develop plans for behavioral action. *Choosing* deals with intellective tasks, such as problem solving when there are correct answers and decision-making tasks when there are no correct answers. *Negotiation* includes tasks aimed at resolving conflicting viewpoints and mixed-motive tasks aimed at resolving conflicts of interest. *Execution* refers to competitive tasks that help to resolve conflicts of power and performance tasks designed to make things or provide services.

For Steiner, the suitability of a task for a group depends on the demands of the task. Task demands vary along three dimensions (Table 2.2). Some tasks are divisible and can be broken down into subtasks for individual group members, whereas other tasks are unitary. Some tasks require a high rate of production (maximization), whereas other tasks require high-quality solutions (optimization). Finally, tasks combine the efforts of group members in different ways. The group's work can be added together, limited by the last member, averaged, selected, or combined in any way the group desires.

TABLE 2.2

Steiner's Demands of Group Tasks

Issue	*Options*	*Task Type*
Can the task be divided?	Subtasks exist	Divisible
	No subtasks exist	Unitary
Is quantity more important than quality?	Quantity	Maximizing
	Quality	Optimizing
How are the individual inputs related to the group's product?	Added together	Additive
	All group members must contribute	Conjunctive
	Average of the individuals	Compensatory
	Select from individual judgments	Disjunctive
	Group decides how to organize	Discretionary

SOURCE: From *Group Dynamics*, 3rd edition, by Forsyth, D. Copyright © 1999. Reprinted with permission of Wadsworth, a division of Thomson Learning, www .thomsonrights.com Fax 800 730–2215.

Additive and conjunctive tasks are usually divisible and maximizing, the subtasks can be identified, and quantity of output is emphasized. Additive tasks combine group member contributions together, such as when a group paints a house. The productivity of a group will exceed that of the individual group member. However, production is often less than the sum of individuals working alone, given that this type of task tends to lower the motivation of individual performers. Conjunctive tasks are not completed until all group members have completed their parts. An example of this is assembly-line work. Although group performance is limited by the worst-performing member, the group compensates in several ways. It can encourage the poor performer to work harder, it can provide advice and support to the poor performer, or it can assign the poor performer to the easiest part of the task.

A compensatory or disjunctive task is typically unitary and optimizing, the task is not easily dividable into subtasks, and quality of output is emphasized. A compensatory task averages the input of group members to create a single solution. An example of this type of task is when the group leader asks for all group members' opinions and forms a single recommendation from their responses. The average score is better than the opinions of most of the individual members. In a disjunctive task, the group must generate a single solution that represents the group's product. Typically, the group discusses the issue until members agree on a solution. The decisions of juries and the problem-solving actions of technical teams are examples of disjunctive tasks. A group usually performs better than individuals in this type of task, but not better than the best individual in the group. The problem in making a correct group decision is how to evaluate the opinions of group members. To be successful, someone must generate the right answer, and the group must decide to adopt it.

When the group is able to decide how it wants to perform a task, it is called a discretionary task. Discretionary tasks may be divisible or unitary, as well as maximizing or optimizing. Self-managing teams are an example of groups that work on discretionary tasks because they are able to decide how to best perform their tasks. The performance of a group on a discretionary task is variable, given that it depends on whether the group selected an appropriate method to perform the task.

Both the McGrath and Steiner systems for classifying team tasks are useful for understanding what a team does. McGrath's system explains the different types of tasks a team actually performs. A team may perform only one or two types of tasks, such as a factory team that primarily performs a physical task and might do some problem solving. Other types of teams may perform many different types of tasks, such as a project team that both designs and produces a product. Understanding the range of tasks performed by a team is important in selecting and training team members.

Steiner's system acknowledges that a team performs a variety of tasks that can be combined in different ways. The team's work may be organized as an additive task (where each person's job is relatively independent) or as a conjunctive task (where team members shift roles to keep the process going smoothly). Steiner's work has been useful in explaining the benefits of and problems with different ways of combining tasks. For some types of tasks, organizing work into teams can create synergies that improve performance over that of individuals. However, using teams may also reduce performance because of coordination and motivation problems. (These performance losses are discussed in more detail in Chapters 4 and 9.)

Group Process

Having the right people and the right type of task for a group does not guarantee success. Team members must be able to combine efforts success- fully. Teams may not reach their potential if their internal processes interfere with their success. Effective teams organize themselves to perform tasks, develop social relations to support their operations, and assign leaders who can provide direction and facilitate team operations.

Groups primarily engage in two types of activities: making decisions and performing tasks. For both of these activities, internal group processes may limit success. Groups may encounter problems with decision making. Groups are imperfect decision makers and do not always fully use their collected knowledge and skills. Group decisions may be disrupted by personal bias, dis- torted by the desire to maintain good relationships, or impaired by the desire to make decisions quickly. Groups often become prematurely committed to the first acceptable solution instead of taking a structured approach to prob- lem solving.

Even when groups are organized for the sole purpose of performing certain tasks, group process issues may have both positive and negative impacts on performance. Highly effective groups have task-oriented goals and norms. These groups outperform collections of individuals, but things may go wrong. A group might have unclear goals or norms that do not encourage perfor- mance of its task. Working in a group may lead to reduced effort by individ- ual members, rather than encouraging performance. (This problem, called social loafing, is discussed in Chapter 4.)

Internal social relations should provide support for the group. Group members must communicate well, work cooperatively together, and provide emotional support for one another. Groups with high levels of group cohesion and good social relations are the most effective groups. If a group is riddled with conflict and divided into cliques, or acts competitively rather than cooperatively, communication can break down.

It is the leader's responsibility to provide direction for the group and facilitate its internal processes. There is no set of rules a good leader can mechanically follow. Depending on their tasks and maturity, groups require different types of leadership. The use of teams often changes the nature of leadership because team leaders do not have the same power and authority as traditional managers.

Organizational Context

The organizational context has a significant effect on whether teams operate successfully (Guzzo & Dickson, 1996). Teams may be used to improve the operations of organizations, but teams are sensitive to their organizational environments and need the right conditions to be successful. The organizational context relates to the culture of the organization, the support it provides for teams, and its evaluation and reward systems.

Teams are more likely to be successful in organizations with supportive organizational cultures. Supportive cultures encourage open communication and collaborative effort. Power and responsibility are given to teams so that they can control their own actions. Although the use of teams can help change an organization's culture, it can be difficult to initiate change when limits are imposed by the existing culture.

A number of organizational supports should be provided to help make teams function more effectively (Hackman, 1990b). Teams perform better when they have clear goals and well-defined tasks. They must be provided with adequate resources, including financial, staffing, and training support. Reliable information from the organization is required for teams to make decisions, coordinate their efforts with other parts of the organization, and plan for future changes. Finally, teams should have available to them technical and group process assistance. They need technical help to solve their problems and facilitation or coaching to deal with interpersonal difficulties.

Building effective teams requires the efforts of both team members and the organization. To improve on the way team members operate, a team needs feedback on its performance and an incentive to change. To an extent, the team can evaluate itself, and team members can provide support for one another, but an effective team requires feedback from the organization and rewards for good performance. Without this, team members will not focus on the goals the organization has established for the team.

3. Characteristics of Successful Teams

What makes for successful teams? Several researchers have tried to answer this question. Their typical approach is to find examples of successful teams and use interviews and surveys to determine what makes these teams successful. Although research approaches are similar, the types of teams researchers investigate often are different. In addition, the types of questions used to examine teams differ depending on the backgrounds of the researchers. Several approaches to defining characteristics of successful teams illustrate the characteristics successful teams have in common.

Hackman (1987) is an organizational psychologist whose specialty is job design. His research has examined a wide variety of teams both at work and in the laboratory. He lists the following five factors as necessary for the successful development of teams.

1. Clear direction and goals: Teams need goals to focus efforts and evaluate performance.

2. Good leadership: Leaders are needed to help manage the internal and external relations of teams and orient teams toward their goals.

3. Tasks that are suited for teamwork: Tasks should be complex, important, and challenging, requiring the integrated efforts of team members, and the tasks should not be capable of being performed by individuals.

4. Necessary resources to perform tasks: These include material, training, and personnel resources.

5. Supportive organizational environment: Organizations must allocate sufficient power and authority to allow team members to make and implement decisions.

Levi and Slem (1995) are psychologists who examined teamwork in high-tech companies. They studied factory production teams and engineering research and development teams to determine factors related to team success. They found the following five factors.

1. Evaluation and rewards: Teams need fair and objective criteria for evaluation, team member performance evaluations should relate to their team contributions, and members should be rewarded when their teams are successful.

2. Social relations: Teams need training in social skills so they can resolve internal conflicts and function smoothly.

3. Organizational support: Management, the organizational system, and the organizational culture must support the use of teams.

4. Task characteristics: Teams need clear direction and goals, tasks that are appropriate for teamwork, and work that is challenging and important.

5. Leadership: Leaders need to facilitate team interactions and provide assistance to teams when problems occur.

Larson and LaFasto (1989) are experts in group communication. They studied a variety of teams from business, sports, and government. As in the previous studies, they found clear goals with standards of excellence, principled leadership, and external support and recognition to be important factors in successful teams. In addition, their research indicated that a results-oriented structure, competent team members, a unified commitment, and a collaborative climate were important.

Katzenbach and Smith (1993) are management experts who studied upper-level management teams, primarily in large organizations. They found that clear performance goals, common approaches and methods for completing tasks, and a sense of mutual accountability were factors related to success. In addition, they believe that a team performs best when there is a small number of team members, members have adequate levels of complementary skills, and there is commitment to a common purpose.

Table 2.3 presents the characteristics of successful teams listed by most of the preceding researchers. Teams require clear, well-defined goals to provide direction and motivation and to allow for performance evaluation. Leaders keep teams focused on goals and facilitate but do not control teams' activities. The organization's cultures and systems must be compatible with teamwork, and organizations must supply teams with the necessary power and resources (e.g., personnel, financial, training) for task performance. Tasks must be suitable for teamwork. Tasks should require coordinated effort and be both challenging and motivating. Finally, team members should have a sense of common fate or mutual accountability, and their efforts must be evaluated and rewarded in a fair manner.

Although it is useful to try to determine the factors that characterize successful teams, these approaches are limited. The differences found by researchers reflect both different research approaches and different types of teams studied. Cohen and Bailey (1997) conducted a meta-analysis of work teams during the 1990s. Their review of 54 studies of work teams shows that the factors important for success are different for production, professional, and managerial teams. For example, for self-managing production teams, the amount of organizational support is very important, but the quality of leadership is relatively unimportant. On the other hand, professional project teams often are dependent on high-quality leadership because of the nonroutine nature of their tasks.

TABLE 2.3

Characteristics of Successful Teams

	Hackman	Levi & Slem	Larson & LaFasto	Katzenbach & Smith
Clear goals	X		X	X
Appropriate leadership	X	X	X	
Organizational support	X	X	X	
Suitable tasks	X	X		X
Accountability & rewards		X	X	X

Summary

The definition of team success relates to team tasks, social relations, and impact on their members. Successful teams complete their tasks and do so in a collective way that is better than using only individuals to perform the tasks. Teams must develop good social relations to support task activities and maintain the existence of the team. Participating in teams should be a benefit to team members, both in terms of learning new skills and advancing individual careers.

The success of a team depends on the composition of the group, the task, the group process, and the organizational context. There are three important aspects of group composition. A group must have members with the right set of knowledge, skills, and abilities to complete its task. For some types of groups, members must represent the relevant parts of an organization to ensure a sense of participation in the decision and support for its implementation. Finally, group members must have the interpersonal skills to work together as a team.

Teams perform a variety of tasks that require different sets of skills. The tasks that groups perform may be analyzed by examining whether they can be divided, whether they focus on quantity or quality, and how member input relates to the products. For a task in which the group's work is simply added together, the group performs no better than the same number of individuals working alone. When the poorest performing member limits completion of a task, the group performs better because it can compensate for individual problems. For a task that requires the group to make a quality decision, the group often performs better than individuals working alone. When the group is allowed to decide how to perform a task, success depends on how well the group structures the task.

The group process connects the members of a group to its task. Successful group processes organize the group to complete the task, develop supportive social relations, and assign leaders to provide direction and facilitation. For each of these steps, the group must overcome obstacles that interfere with its interpersonal dynamics.

The organization provides a context for the team. The organization's culture supports the team by creating an environment that encourages collaboration and allows the team to control its internal operations. The organization's systems support the team by providing direction, resources, information, and assistance. One of the most important aspects of the organizational context is the willingness of the organization to provide feedback on a team's performance and to reward successful performance.

Researchers from a variety of perspectives have identified several common features of successful teams. Teams have clear goals that provide direction and motivation. Their leaders structure tasks and facilitate group processes. Their organizations provide supportive contexts for the teams to grow. The tasks that teams perform are well suited for teamwork. Finally, team members are held mutually accountable for the success of their teams, and they are rewarded for their efforts. Although they are characteristics of successful teams, the importance of these various characteristics changes depending on the types of teams.

TEAM LEADER'S CHALLENGE—2

You are the team leader of a quality improvement team for your school district. For the last several months, the eight high school teachers on the team have analyzed problems and developed recommendations for improving how the school system operates. You are proud of the team. They responded well to teamwork training and learned how to operate effectively as a team. In addition, social relations among the teachers on the team have been very good and several strong friendships have developed.

You presented the recommendations developed by the team to the superintendent of the school district. After waiting for several weeks, the superintendent thanked you for your efforts, but told you that none of the team's recommendations would be implemented at this time because of budget constraints. You feel rejected by the superintendent and discouraged. It is time for your last team meeting. You need to prepare what you are going to tell the team.

How should the team leader handle the last meeting with the team?

In what ways was the team successful and unsuccessful?

How can the organization better use improvement teams in the future?

ACTIVITY: UNDERSTANDING TEAM SUCCESS

Objective. Why are some teams successful while others are unsuccessful? This activity tries to use your experience with teams to answer this question.

Activity. Think about a time when you were on a successful team. Using Activity Worksheet 2.1, write a description of the team at that point in time. (What was it like being on the team? What was the team like?) Think about a time when you were on an unsuccessful team. Write a description of the team at that point in time.

ACTIVITY WORKSHEET 2.1

Successful and Unsuccessful Teams

Successful Team:
Unsuccessful Team:

Analysis. Compare the two descriptions of successful and unsuccessful teams. What characteristics may explain the differences between these two teams? Compare your answers with those of other group members. Are the characteristics similar? Develop a group answer to the following question: What are the characteristics of successful teams?

1. _____

2. _____

3. _____

4. _____

Discussion. Using your list of the characteristics of successful teams, what advice would you give a team leader about how to establish and run a team?

PART II

Processes of Teamwork

3

Team Beginnings

Teams develop through a series of stages reflecting changes in internal group processes and the demands of their tasks. One of the most important insights of the stage perspective is that teams often are not productive at the beginnings of projects.

To become more effective, teams should address several issues when first formed. The team should orient or socialize new members into the group; this socialization process assimilates new members while accommodating their individual needs. The purpose or objective of the team should be defined by creating team goals. Developing team goals is an important process that helps avoid problems, provides directions, and increases motivation. The team should develop rules or group norms by which to operate. These norms define appropriate behavior for group members.

There are techniques to help teams form social relations, clarify the definitions of tasks, develop team norms, and create a team contract. If these issues are focused on at the beginning, teams are more likely to be effective.

Learning Objectives

1. Understand the main stages of group development.

2. How do the demands of a project change the way a team operates?

3. What are the implications of team development stages?

4. How do evaluation, commitment, and role transitions affect group socialization?

5. What are the main characteristics of team goals?

6. What are hidden agendas? How do they affect a group?

7. What are the main functions of group norms?

8. What are the positive and negative effects of group norms?

9. What can a team do to help improve the beginning stages of a team project?

1. Stages of Teamwork

Research on project teams shows that the start-up activities take longer than anticipated. For many professional design projects, most of the design work occurs during the last half of the allotted time (Gersick, 1988). The main reason for this slow start is that it takes time to decide on the definition and goals of a project, to develop the social relations, and to create effective operating rules. An understanding of the dynamics of team projects and group development can help speed up this process and reduce frustration caused by what members often perceive as a slow start to a team's work. Teamwork is not a smooth process; a team goes through stages and has its good and bad periods. It is important to develop the capabilities of the team while working on its task.

A variety of theories has been created to explain the changes that groups experience during their existence. Stage theories of group development focus on how internal group processes change. Project theories attempt to describe how groups change based on the tasks that they perform. Finally, alternative theories explain group process changes as cycles rather than stages.

Group Development Perspective

There are many stage theories of group development. However, most of the theories have similar elements. The theories try to explain why it takes time for a group to develop before it becomes productive, and why the group goes through periods of conflict during its development. Table 3.1 presents the best-known group development stage theory, developed by Tuckman and Jensen (1977). This theory focuses on the development of internal relations among the team members.

A group begins with the *forming* stage, where few measurable accomplishments occur. During this stage, group members get to know one another and learn how to operate as a group. Members tend to be polite and tentative with one another and compliant toward the leader. They often feel uncomfortable and constrained because they are unfamiliar with the other

TABLE 3.1

Stages of Group Development

Stage	Activity
Forming	Orientation: members getting to know one another
Storming	Conflict: disagreement about roles and procedures
Norming	Structure: establishment of rules and social relationships
Performing	Work: focus on completing the task
Adjourning	Dissolution: completion of task and end of the group

SOURCE: From Tuckman, B., & Jensen, M., Stages of small group development revisited, in *Group and Organizational Studies, 2,* copyright © 1977. Reprinted with permission of SAGE Publications, Inc.

members. Group members are confused and uncertain about how to act, and they need to spend time defining their goals and planning how to do their tasks. This stage ends when the group has reached a level of familiarity such that members are comfortable interacting with one another.

The *storming* stage that follows often is characterized by conflicts among group members and confusion about group roles and project requirements. Disagreements over procedures may lead to expressions of dissatisfaction and hostility. Group members may begin to realize that the project is more difficult than anticipated, and they may become anxious, defensive, and accusatory. Members may become polarized into subgroups as conflict about their roles and their views of the task expands. Although this conflict might be unpleasant, it is important because it promotes the sharing of different perspectives and leads to a deeper understanding of members' positions. Not only does this conflict clarify the group's goals, but its resolution leads to increased group cohesion.

The group begins to organize itself to work on the task during the *norming* stage. Here, the group becomes more cohesive, conflict is reduced, and team confidence improves. The group has established some ground rules (or norms) to help members work together, and social relations have developed enough to create a group identity. Increased levels of trust and support characterize group interactions. Although differences still arise, they are handled through constructive discussion and negotiation.

Next is the *performing* stage. The group has matured and knows how to operate, so it now can successfully focus on its task. If the group has developed norms and successfully built social relations, it can easily handle the stress of approaching deadlines. The group focuses on performance through collective decision making and cooperation. Studies of groups show that most performance occurs during this stage, near the end of the group project (Hare, 1982). However, not all groups get to this stage; they may get bogged down in the conflict-oriented stages.

The final stage is the *adjourning* or *dissolution* stage. Some groups have planned endings. When these groups complete their tasks, they disband. Groups also end because of failure to accomplish goals or because of unanticipated problems that make continued group interaction impossible. The adjourning stage can be stressful to group members because they are ending the social relations that they have developed.

During these termination periods, teams should spend time evaluating their performance and using this feedback to prepare for the future (Wheelan, 2005). Debriefing sessions should review the processes and methods of the team. However, team endings often are not learning experiences for the team; members may be more interested in celebrating their success or finding excuses for their failures than learning from the team experience (Hackman & Wageman, 2005).

Some groups do not have planned endings, such as work teams that are designed to continue in operation indefinitely. Like other types of groups, they go through the initial development stages until they reach the performing stage. However, they do not continue performing at a high level forever. Teams can become unproductive for a variety of reasons (McGrew, Bilotta, & Deeney, 1999). Research on work teams suggests that after a few years, they may arrive at a point when longevity no longer is a benefit to performance (Guzzo & Dickson, 1996).

The Tuckman and Jensen model (1977) has been criticized as being too rigid a stage theory. Alternative models of group development highlight the ongoing, cyclical, and developmental nature of groups. For example, Wheelan's (2005) integrated model of group development recognizes that groups may get stuck in various stages or repeat stages due to problems encountered. Wheelan's developmental model has a similar set of group stages, but she emphasizes the activities that leaders and members may perform to promote group development and effective performance.

Project Development Perspective

An alternative view of group stages is based on the characteristics of projects rather than the development of group processes. These theories are based on research on work teams, whereas group development theories often are based on therapy or learning groups. For example, McIntyre and Salas (1995) present a model of team development based on the skills team members develop while completing a project. In their model, a team works on role clarification during the early stages, moves on to coordinated skills development, and finally focuses on increasing the variety and flexibility of its skills as a team. This model uses the changing relationship between the team and the project as the driver of change throughout the stages.

McGrath (1990) proposes a model of how project groups operate over time. The group performs four types of functions: inception (selecting and accepting goals), problem solving, conflict resolution, and execution. Each of these functions has implications for how the group operates and for relations among group members. During the inception stage, the group is focused on planning activities and collaborating. During the conflict resolution stage, social relations are strained because the group is dealing with conflict. The problem-solving and execution stages focus on coordinating the ideas or actions of the group members.

In McGrath's model, a group does not necessarily need to perform all these functions to achieve its goals. For example, on simple problems the group may go directly from inception to execution without the middle functions. For highly complex problems, the group may have to recycle through the four functions a number of times.

The time available for a project group affects the way in which the group operates. The more time allotted, the more time a group will spend analyzing problems and developing its social relations. Therefore, giving a group sufficient time to complete its tasks tends to lead to better quality projects and a more developed group. Over the long term, a group that is always dealing with crises will tend to produce lower-quality decisions and have poorer group processes.

Ancona and Caldwell (1990) present a model of group development for new product teams. Their three stages of development are based on the changing nature of the tasks a team performs. In addition, rather than focusing solely on the internal processes of the group, the model highlights the changing emphasis on internal and external relations.

During the creation stage, a team's activities are a mixture of internal and external processes. The team is developing new ideas and creative solutions while organizing the work team. External relations include gathering information from the organization, understanding the organizational context of the project, and building links with relevant organizational units. During the development stage, the team's focus is primarily internal. The project's idea has been approved by the organization, and the team is focused on mastering the technical details of the project. Motivation and coordination of internal activities are the central focus. The final stage is diffusion, where external relations become the primary focus of the team. The project is nearly complete, and coordinating its transfer to manufacturing and marketing is the focus of the team's activities. Internal processes often are strained, so the leader has to work on keeping team members together and motivating them to complete the project.

Alternatives to Stage Theories

Although stage theories of group development are popular, not all groups follow these patterns. Some groups skip stages, others get stuck in certain stages, and still others seem to travel through the stages by different routes. The boundaries between the stages are often less clear-cut than the theories suggest.

Rather than emphasizing a sequence of stages, some group theorists believe that groups go through cycles that can be repeated throughout the life of the group. For example, Bales's (1966) equilibrium model of group development views groups as balancing the needs for task completion and relationship development. Depending on the needs of their members, groups go back and forth between these two concerns.

Marks, Mathieu, and Zaccaro (2001) developed a recurring phase model of teamwork showing how teams perform in temporal cycles of activity that create a rhythm for the team. These cycles may vary from the hourly tasks of production and service teams to project teams with multi-year cycles. The length of the cycle is determined by the task. Teams operate in cycles of action (when they performing their tasks) and transition (when they are evaluating their performance and planning for the future).

Gersick (1988) developed a theory of punctuated equilibrium from her research on project teams. Each team had its own pattern of development, but all the teams experienced periods of low activity, followed by bursts of energy and change. In addition, each team had a midpoint crisis in which its members realized that half their time was gone but the project was still in its early stages of completion. This led to a period of panic, followed by increased activity as the team focused on completing the task.

Implications of Team Development Stages

Understanding the stages that teams typically go through can help team members better recognize what is happening to the team and how to manage it. Stage theories explain why most of the team's work gets done at the end of the project, and why it is important to build social relations and team norms at the beginning of the project. However, it is important to remember that stage theories of team development do not always apply. A team's life is often a roller-coaster of successes and failures. Some teams get stuck at one of the stages or even break up; they never get to the performing stage because they have not worked through their earlier problems (Wheelan, 2005).

Several lessons are important here. First, emotional highs and lows are a normal part of group development. Second, developing the group is important. Time must be spent developing social relations and socializing new members, establishing goals and norms, and defining the project. Third, the group may go through periods of lower task performance as it tries to resolve conflicts over relationship and task issues; this is a normal part of group development as well.

2. Group Socialization

Group socialization refers to the process by which a person becomes a member of a group. An individual goes through a series of role transitions—from newcomer to full member—during the socialization process. At each step in this process, the individual is evaluating the group and deciding on his or her level of commitment (Moreland & Levine, 1982). Evaluation is the judgment whether the benefits of participation in a group outweigh the costs. Commitment is the desire to maintain a relationship with the group. These processes are mutual. The individual evaluates the group and decides on a level of commitment, and the group evaluates the individual and decides how committed it is to him or her.

Evaluation and commitment usually are positively related. People who evaluate their groups positively are usually more committed to their groups. However, people may become committed to groups they do not evaluate positively. A work group may be undesirable, but individuals feel committed to it because of the desire to retain their jobs. A highly demanding work group might not be enjoyable, but high levels of involvement by other members will often encourage involvement and commitment.

The socialization process starts with the investigation stage, where each side searches for information. The group attempts to recruit the individual, while the individual engages in reconnaissance to decide whether to join the group. This stage ends when the individual decides to join the group.

The socialization stage determines how the individual will be integrated into the group. The individual must learn the norms and practices of the group, and must accept its culture. This process of assimilation is matched by the group's accommodation to fit the newcomer's needs. This stage ends when the individual has been accepted as a full member of the group.

During the socialization stage, the newcomer spends time seeking out what is expected of him or her by the group, and group members provide information through both formal and informal orientation activities (Wanous, 1980). At the beginning, the newcomer often is anxious about his or her role in the group and tends to be passive, dependent, and conforming. This style actually increases the newcomer's acceptance by established group members (Moreland & Levine, 1989). The newcomer is a threat to the team because he or she brings in a fresh and objective perspective that can be unsettling to existing members. The passive approach adopted by many new-comers reduces the potential threat of criticism of the team, and thereby encourages acceptance of the new member.

During the maintenance stage, the individual is fully committed to the group. Even though the individual is a full member of the group, there is an ongoing process of negotiating his or her role and position in the group and the group's goals and practices. Although many members stay in this stage until they leave the group, members may diverge from the group and reduce their commitment because of conflicts between their personal goals and those of the group.

Turnover of team members can have positive and negative effects on a team (Levine & Choi, 2004). The introduction of newcomers requires existing team members to spend time and energy socializing them into the team's operations. However, the newcomer may improve the team's operations by introducing innovations in the way the team operates. A change in team members may improve team performance by adding new perspectives to the team's decision-making processes. To the extent that newcomers stimulate discussion about how the team operates, their introduction may lead to improvements in team performance. In essence, newcomers may encourage the team to rethink how it operates during the process of socializing new members.

3. Team Goals

Group goals are "a desirable state of affairs members intend to bring about through combined efforts" (Zander, 1994, p. 15). A clear understanding of a group's objectives through well-articulated goals is the most common charac-teristic of successful teams (Larson & LaFasto, 1989). Research on work teams shows that clear project goals help to improve team performance and internal team processes (McComb, Green, & Compton, 1999). A team with shared goals is more likely to complete its tasks on time and have less internal conflict.

Goals are only one way a team can define its purposes. Teams often cre-ate mission statements or charters that in general terms define their purposes and values. A mission statement articulates a team's values but does not say how the team's purposes will be fulfilled. A team's goals must be consistent with the mission statement, but goals should be objective, defining the accomplishments that need to be completed for success. A team also can create subgoals or objectives that serve as signposts along the way to completion of main goals.

Group goals can be oriented to different types of performance. They may focus on quantity, speed, quality, or service to others. It is clear that specific goals, regardless of form, improve performance better than absent or ill-defined goals. However, the improvement is directly related to the form of the goals (Guzzo & Dickson, 1996). Goals for quantity improve quantity, but not necessarily other aspects of the group's performance.

Value and Characteristics of Goals

The value of team goals is to provide the team with direction or vision and motivation. Good team goals are clear and specific, so that team members can understand them and relate the goals to their performance (Locke & Latham, 1990). Goals should be moderately difficult; that is, motivating but not impossible to achieve. Team goals work best when the task is interesting and challenging and requires that team members work together to succeed.

Teams perform better when they are able to participate in setting challenging performance goals (Kerr & Tindale, 2004). Participation helps gain acceptance and support for the team's goals. The goal-setting process helps the team to better understand the task. Participation also encourages collective efficacy, or the belief that the team is capable of improving performance and meeting its goals.

Goals serve a variety of functions for a team. Table 3.2 lists several of these functions. Goals help direct and motivate the team and its members, but they also serve functions outside the team; they help establish relationships with other groups and establish criteria for evaluation from the surrounding organization.

Zander (1994) believes that the three most important characteristics of group goals are accessibility, measurability, and difficulty. Accessibility refers to the probability of completing goals. This relates to the group's knowledge of how to complete the task and meet the goals. For example, research and development teams might set goals that turn out to be impossible to meet, or management teams might not know how to implement a task once the goals are formed. These situations are quite different from that of a production team needing a little guidance and a lot of motivation to complete its task.

TABLE 3.2

Functions of Team Goals

---

1. Serve as a standard that can be used to evaluate performance

2. Motivate team members by encouraging their involvement in the task

3. Guide the team toward certain activities and encourage integration of team members' tasks

4. Provide a criterion for evaluating whether certain actions and decisions are appropriate

5. Serve as a way to inform outside groups about the team and establish relationships with them

6. Determine when team members should be rewarded or punished for their performance

---

SOURCE: Adapted from Zander, A., *Making groups effective.* Copyright © 1994 by Jossey-Bass.

Measurability refers to the ability to quantify whether the team is reaching its goals. If progress toward the goals is not measurable, the team cannot receive adequate feedback on its performance. Research on process improvement teams shows that measurable goals are necessary for a team to improve its performance. Without measurability, one might as well tell the team to "go out and do your best"—a nice motivational expression that does not lead to improved performance.

The final characteristic of good goals is difficulty. The goals need to be at least moderately difficult to encourage and motivate performance so that the team feels a sense of accomplishment when it reaches the goals. However, care should be taken not to establish goals that are too difficult. When a team repeatedly misses its goals, members become embarrassed, begin to blame one another and outside factors for performance problems, and may refuse to commit to goals in the future (Zander, 1977).

It is important to note that a team is not always free to set its own goals. Teams have relationships with outside groups, such as customers, that depend on the team's efforts and that influence its goals. Teams also set goals relative to the performance of comparison groups such as other organizations. Finally, team goals are strongly influenced by their organizational and managerial contexts. With the exception of top management teams, the goals of most teams are defined by their organizations.

Although goals may play an important role in providing direction and motivation for a team, this is not invariable. Often a team is faced with a situation in which the goals are unclear or team members have differing views of the goals. For a project team, sometimes understanding the problem it is trying to solve (i.e., defining the goals of the project) is more difficult and time-consuming than developing a solution. Improving the quality of team goals through goal-setting activities is a solution to these problems and is discussed in Chapter 17.

Hidden Agendas

Group goals provide a number of valuable functions, but they also can be a source of problems. Problems arise from hidden agendas, which are unspoken individual goals that conflict with overall group goals.

The most basic type of hidden agenda relates to the motivational aspect of goals. Although the team may decide to commit itself 100% to doing a high-quality job on a project, some team members may not believe the team's activities are important. They might decide to slack off and spend more time and effort on other activities. Their goal is to help the team succeed with the least amount of effort on their part.

A second type of hidden agenda relates to the directional aspect of goals. Some team members may not agree with the goals of the team, or they may have individual goals that are incompatible with the team goals. For example, in an organization budget committee, team members must deal with the potentially conflicting goals of doing what is best for the organization versus doing what is best for the departments they represent.

These differences between individual and group goals can create conflict within a group. It is a difficult conflict to resolve because it is hidden. A low-motivated team member will create excuses rather than telling the team he or she is unwilling to work hard on the project. A team member with conflicting loyalties will hide this conflict, leading other team members to distrust what is being said. The overall effect of hidden agendas is to damage trust within the team. Over time, this reduces communication, creates conflicts, and makes conflicts more difficult to resolve.

By the time the team is aware that a hidden agenda is its problem, it probably already is suffering from conflicts and distrust. Directly confronting people about hidden agendas often does not work; it just forces them to be defensive and deny that there is a problem. Hidden agendas often are better dealt with indirectly (Johnson & Johnson, 1997). Rather than confronting a team member directly about a hidden agenda, the team can strengthen its group goals or improve its communication processes.

Focusing on group goals is an indirect approach to dealing with hidden agendas. Can the group adopt a set of goals that team members can agree to support? Spending time evaluating and modifying the group's goals so that they are acceptable to all members reduces the possibility of a hidden agenda. An alternative approach is to deal with the communication problems created by hidden agendas. Conducting group activities that encourage honest communication and build trust among team members is useful. Open and trustworthy communication enables the team to better manage its conflicts and may allow team members to safely discuss goal conflicts.

4. Group Norms

Group norms are the ground rules that define appropriate and inappropriate behavior in a group. They establish expectations about how group members are to behave. These rules may be explicit (e.g., 51% majority required on all votes for a decision) or implicit (e.g., group members take turns when talking). Although most groups do not formally state their norms, members typically are aware of the rules and follow them.

There are four main functions of group norms (Feldman, 1984). First, group norms express the group's central values, which help give members a sense of who they are as a group. Second, norms help coordinate the activities of group members by establishing common ground and making behavior more predictable. Third, norms help define appropriate behavior for group members, allowing members to avoid embarrassing or difficult situations and thereby encouraging active participation in the group. Fourth, norms help the group survive by creating a distinctive identity; this identity helps group members understand how they are different from others and provides criteria for evaluating deviant behavior within the group.

A number of factors affect the power of group norms to control the behavior of its members (Shaw, 1981). The more clear and specific a norm is, the more members will conform to it. If most group members accept and conform to the norms, others are more likely to conform. The more cohesive a group is, the more conformity there is to group norms. Groups are more tolerate of deviance from peripheral norms than from norms that are central to their operations (Schein, 1988). For example, technical experts who are valuable contributors to the team may be allowed to violate peripheral norms concerning dress codes or rules of social etiquette.

How Norms Are Formed

Group norms often develop unconsciously and gradually over time. They are created by mutual influence and develop through the interactions of group members. Even though people obey these norms, they may be unable to articulate them. In addition, group members typically obey norms even when there is no external pressure, such as threats of punishment, to do so. This shows that they have accepted the norms and are using them to guide their own behavior.

Group norms come from a variety of sources. Groups can develop norms based on those from other groups to which they have belonged. Norms can be based on outside standards such as those by which other social or organizational groups operate. Norms also are strongly influenced by what happens early in the group's existence, and they are most likely to develop in situations in which members are unsure how to behave. For example, when a group is having problems with members showing up late for meetings, the group is likely to develop explicit norms for attendance.

Many teams simply ignore the notion of group norms. They assume that everyone knows how to behave in a group and that there is no need to take the time to create explicit norms. It is not until a team starts to have problems that it becomes aware that members are operating under different norms. Teams benefit from discussing and establishing explicit group norms, which prevents the development of inappropriate norms (e.g., it is acceptable to be late in submitting one's part of the project) and makes everyone aware of the behaviors that are expected. Table 3.3 presents some issues to consider when establishing group norms for team meetings.

Impact of Group Norms

Group norms have positive and negative aspects. Because they control the group's interactions, norms allow fairer communication, maintain respect among members, and distribute power to weaker members of the group. Strong norms are a benefit to the internal workings of the group. However, norms also enforce conformity, which can be a problem from the organization's perspective.

The Hawthorne studies of teamwork showed the benefits of and problems with group norms. In this factory setting, group norms controlled the amount of work people performed. When the team had high performance norms, norms were a benefit because they kept laggards in line and encouraged workers to help one another. When the team had low performance norms, however, the ability of management to change the team's behavior was limited because group norms were resistant to outside influence.

TABLE 3.3

Norm Issues for Team Meetings

Decisions	How should decisions be made? Must everyone agree for consensus? Does anyone have veto power?
Attendance	What are legitimate reasons for missing meetings? How should the team encourage regular attendance?
Assignments	When assignments are made, what should be done when team members do not complete them, or complete them poorly?
Participation	What should be done to encourage everyone to participate?
Meeting times	When do meetings occur? How often should the team meet? How long is a team meeting?
Agendas & minutes	Who is responsible for these activities? What other roles should be set up?
Promptness	What should be done to encourage promptness?
Conversational courtesies	How can the team encourage members to listen attentively and respectfully to others? Does the team need rules to limit interruptions or prevent personal criticisms?
Enforcement	How should the team enforce its rules?

SOURCE: Adapted from Scholtes, P., *The team handbook: How to use teams to improve quality*. Madison, WI: Joiner Associates. Copyright © 1988.

5. Application: Jump-Starting Project Teams

Organizing people into teams to complete projects often leads to initial drops in performance (Katzenbach & Smith, 1993). It takes time for groups to develop their internal social processes. Teams that have been working together often are more productive than new teams. Part of this difference relates to the shared knowledge that develops among team members, known as "transactive memory systems" (Wegner, 1986). Team members develop a shared knowledge base about how to perform a task. Each team member learns the knowledge and skills of the other team members by working together on a project. This shared knowledge improves coordination of activities and increases team performance (Moreland, Argote, & Krishnan, 1996).

When new teams are formed to complete projects, techniques can be used to help speed their development. Teams can learn to operate more smoothly; this experience helps them better manage their projects, improves performance, and reduces the stress of finishing projects in a hurry. Improving teamwork requires effort at the beginning of the project; teams that start off well often perform better as time goes on (Hackman, 1990a). This is why spending time designing and launching a new team is important. The aim is to improve social relations, better define projects, and create a team contract that defines goals, roles, and norms.

Team Warm-Ups

One of the problems with teams is the tendency to focus almost exclusively on tasks. It is equally important to recognize the value of developing a team's social relations. Many of the models for stages of group development state that social relations precede the group's performance stage. Developing social relations among team members also aids in socializing new team members; thus, it is important to focus on developing social relations early in the team's existence.

Team warm-ups are social "icebreakers" conducted at the start of team meetings (Scholtes, 1988). They are crucial to first team meetings and should be used during the early stages of team formation to develop social relations within the team. Warm-ups are social activities designed to help team members get to know one another and improve communication during the team project. Warm-ups can be as simple as spending five minutes sharing favorite jokes or chatting about what team members did over the weekend. Table 3.4 contains several common team warm-up exercises that are useful during the early stages of a team's life.

Project Definitions

Teams often jump into projects and then have to back up to earlier stages when problems arise. In the rush toward task completion, a team may have spent too little time understanding the assignment (Pokras, 1995). Everyone on the team should have the same understanding of the assignment, and this understanding should conform to the organization's understanding. Many professional teams find the project definition stage to be the most difficult and important stage.

There are many reasons why teams try to skip over the project definition stage. Team members may feel socially uncomfortable at the beginning,

TABLE 3.4

Team Warm-Up Exercises

1. Team Introductions	Team members take turns introducing themselves. In addition to giving their name, team members may tell something about themselves, such as where they grew up, their favorite joke, favorite songs and performers, or favorite movie. This exercise can be repeated at later team meetings by adding more information about each team member.
2. Team Name	Team members each write down five names they think would be a good team name. Each team member then states his or her five names. The team then creates a new list of team names that does not overlap with the existing list. The goal is to select a name for the team from the second list. All team members should support the selected name.
3. Conversation Starters	It is often useful to spend 5 or 10 minutes having a group discussion about social and personal topics. The following list contains some useful conversation starters: • What types of work activities do you like and dislike? • What things do you like most and least about team projects? • What values are most important to you? • What do you do when you want to relax? • What kinds of things make you feel uncomfortable?

SOURCE: Adapted from Scholtes, P. (1988). *The team handbook: How to use teams to improve quality*. Madison, WI: Joiner Associates. Copyright © 1988.

so they want to quickly focus on performing the task. Task assignments are often ambiguous, and this ambiguity causes discomfort; making quick decisions to clear up the problem is emotionally satisfying. Such actions may help address the emotional aspects of the problem, but they often result in a team heading off in the wrong direction, leading to conflict and time delays later in the project.

Teams can use several techniques to improve their ability to define a problem and understand its underlying causes. These are discussed in Chapter 11.

The team should use these techniques at the beginning of the project to better understand its assignment.

Team Contract

A proper team launch includes developing common team goals and objectives, clarifying roles, creating appropriate team norms, and defining performance expectations. An effective way to do this is to develop a team contract or charter (Herrenkohl, 2004). The contract explicitly states the agreements the team has reached on how to operate. The act of developing a contract helps the team identify and resolve conflicts and misunderstandings. It is a valuable technique for getting started in the right direction. Table 3.5 presents an outline of the issues to be included in a team contract.

TABLE 3.5

Creating a Team Contract

---

1. *TEAM GOALS:* What are the main goals of the team?

2. *TEAM OBJECTIVES:* What specific actions relate to the team's goals? How can these actions be measured or evaluated?

3. *TEAM MEMBER ROLES:* What are the primary roles and responsibilities of each team member?

4. *TEAM NORMS:* What norms or operating rules should the team develop? Discuss norms for decision making, attendance, performing assignments, participation, and meetings. How should the team enforce its norms?

5. *TEAM MEMBER EVALUATION:* What criteria will be used to evaluate each team member's performance?

Summary

Groups develop through a series of stages from formation to adjournment. These stages relate to the time needed to develop a group's internal processes and the changing demands of its tasks. The developmental perspective shows the different types of challenges groups face during their existence. Rather than a smooth progression, groups go through periods of low activity followed by bursts of achievement, and from periods of smooth relations to conflict. Understanding these stage theories helps explain why groups do most of their productive work during the later stages of projects.

The group socialization process describes the changing relationship between a group and its members. The group and its members evaluate one another to determine reciprocal levels of commitment. Socialization proceeds through a series of stages, from investigation to maintenance. Each stage is marked by a role transition that reflects the changing level of group members' commitment.

Goals define a group's purpose and values and are an important factor in the group's success. Goals often are divided into objectives that are linked to performance criteria. Effective group goals are accessible to the group, measurable in such a way as to provide feedback on performance, and moderately difficult to achieve in order to motivate performance. One common goal problem for the group is hidden agendas. Hidden agendas occur when individual group members have unspoken goals that conflict with the group goals. These can create conflict and distrust in the group and must be managed carefully.

Group norms define appropriate behavior for group members. They help the group operate more smoothly and create a distinctive group identity. Norms often evolve gradually; however, a group should formally establish its operating norms. The impact of group norms can be both positive and negative. Norms help a group operate better internally, but they may make the group more resistant to change by the outside organization.

One of the values of viewing the ways groups change over time is that it illustrates the problems teams have at the beginning of projects. Teams must address the problems of undeveloped social relations, ill-defined projects, and ambiguous goals and norms before they can focus on performing their tasks. The operation of teams can be improved by focusing on these problems at the beginning of team projects.

TEAM LEADER'S CHALLENGE—3

It is the first meeting of a new product development team. The team's goal is to create the next generation of kitchen appliances for the company. The team contains members from engineering, product design, marketing, manufacturing, and finance. This project will be the main work activity for team members for the next 8 to 10 months.

As the team leader, you need to quickly get the team started on the project. Management has given you an overall goal for the project, but you have had limited time to plan how to manage it. Because of the limited time for completing the project, you are concerned about getting the team off to a good start.

What are the three most important issues for the team leader to focus on at the beginning of the team project?

How much project planning should the team leader do before meeting with the team?

What should happen at the first meeting?

ACTIVITY: OBSERVING TEAM NORMS

Objective. Norms define the rules for appropriate and inappropriate behavior. Although team members often follow norms, most teams do not develop a formal set of norms. Teams may have norms for a variety of issues. Norms may be enforced by official sanctions (e.g., a fine for a violation) or by informal pressure from the leader or team members.

Activity. Observe a team meeting or a group discussion of the Team Leader Challenge. Using Activity Worksheet 3.1, note the norms being used to make decisions, manage participation, and encourage conversational etiquette. For example, does the team use voting to make decisions? Is everyone required to participate before a decision is made? Are there rules to prevent people from interrupting each other?

An alternative norms activity (suggested by Burn, 2004) is to have four-person groups play a card game. After playing for one-half hour, they should identify the formal and informal norms that are operating.

Analysis. After developing a set of norms that the team is using, note how well the team follows them. Do the team members consistently follow norms? Are there examples of people violating norms? How does the team respond to violations?

Discussion. Does the team you observed have effective norms? Are the norms explicit or implicit? If you were asked to provide advice to the team, would you recommend that it develop formal norms? What norms do you think the team should formally adopt? Why?

ACTIVITY WORKSHEET 3.1

Observing Team Norms

Decision-Making Norms
Participation Norms
Conversation Etiquette Norms

4

Understanding the
Basic Team Processes

Motivation, group cohesion, role assignments, and performing both task and social behaviors are the basic building blocks of successful team performance. Working in a team should help motivate its members, but often individual effort decreases when performed in a group. This phenomenon is called social loafing. Developing challenging tasks that require interdependent actions, improving the reward system, developing motivating goals, and increasing commitment to the group can help reduce social loafing and motivate the group.

Beyond motivating a group, successful performance depends on other factors. Group cohesion is the bond that ties the members together. Cohesive groups generally perform better, but cohesion also can cause performance problems. Like the roles in a play, people perform roles in a group. Poorly defined roles can lead to stress, while clear roles help groups operate more efficiently. Although task behaviors typically dominate in work teams, social behaviors are necessary to build relationships among group members. Groups sometimes suffer from a lack of activities aimed at building relationships among members.

Learning Objectives

1. What factors cause social loafing in a group? How can social loafing be prevented?

2. What encourages a group to be more motivated?

3. What factors encourage group cohesion?

4. How does cohesion affect group performance?

5. What are the causes of role ambiguity and role stress?

6. What are the formal roles commonly played by team members?

7. What are the main tasks and social behaviors performed by a group?

8. Why is it important for a team to enact social behaviors?

9. What is the value of group process observations?

1. Motivation

The potential of teamwork lies in the fact that a whole is greater than the sum of its parts; the collective work of a group of people is more than its individuals could accomplish separately. However, although group synergies, the creativity in conflicting ideas, and the motivating impact of team spirit should give a team an advantage over a collection of individuals, it does not always work out this way. In some circumstances, working together causes a decrease in motivation that may be due to social loafing. Understanding this motivation problem suggests the ways teams can increase group motivation.

Social Loafing

One of the biggest motivation problems for teams is social loafing, which is the reduction of individual contributions when people work in groups rather than alone (Latane, Williams, & Harkins, 1979). A simple experiment will demonstrate social loafing. Record the volume when people are asked to shout as loud as they can when they are alone. When asked to do the same task in pairs, the volume will be 66% of two individuals shouting alone. When asked to perform in six-person groups, the volume will be 36% of six individuals.

Social loafing is related to several other group phenomena. People can become "free riders" who perform little in a group because they do not believe their individual efforts are important, and they know they will receive their share of the group's reward regardless of their efforts (Sweeney, 1973). The "sucker effect" (Johnson & Johnson, 1997) is when good performers slack off in teams because they do not want others to take advantage of them. This can lead to all group members reducing their contributions to the task.

A variety of factors contribute to social loafing (Karau & Williams, 1993). If the tasks the team is performing are just a collection of individual tasks, there might be no need for the team to perform in a coordinated way. This may reduce motivation because of the lack of a perceived need to work as a team. Individual performance can be hidden in the team's collective effort, leading members to reduce their effort because they are no longer concerned about what others think of their performance. Finally, team members might be unaware of how much effort others are putting into the task. As a result, they do not know whether they are doing their fair share. Unfortunately, people tend to overestimate the extent of their contributions to the group.

One of the best ways to understand social loafing is to look at a situation where it rarely occurs—a championship basketball game. Only the team's score counts in determining the winner, but everyone's individual participation is observable and measurable. The task is motivating in itself and becomes more motivating through the social aspects of performance. The task requires an integrated and coordinated performance. One player cannot win the game by him- or herself, so each player is dependent on the coordinated efforts of the team to win. Winning is important, and success is highly rewarded. There is no social loafing in basketball or in other tasks that share these characteristics.

Research on work groups shows that these sports principles apply to work. When work teams are given challenging tasks, when they are rewarded for group success yet have identifiable individual performance indicators, and when there is commitment to the team, social loafing does not occur (Hackman, 1986).

Increasing Group Motivation

The discussion of the impact of social loafing on a group helps identify the factors that encourage motivation in the group. Increasing a team's motivation depends on the task it performs, how performance will be evaluated and rewarded, the goals of the team, and the team members' sense of commitment or belonging.

Task. A team is more motivated when the task it performs is interesting, involving, and challenging. Probably the best description of how to create this type of task comes from the job characteristic model (Hackman & Oldham, 1980). A satisfying job creates three critical psychological states: experienced meaningfulness, responsibility for outcomes, and knowledge of results. A task is meaningful when it provides the opportunity to use a variety of skills, to complete an entire piece of work from beginning to end, and to affect others with its completion. Responsibility is experienced when given autonomy or the freedom to design, schedule, and carry out the task as desired. Knowledge of results comes from feedback on the effectiveness of one's performance.

However, a good team task is more than just a good individual task. A good team task requires task interdependence; team members must work together to successfully complete the task. Task interdependence is an additional factor that can be added to the job characteristic model (Van der Vegt, Emans, & Van de Vliert, 1998). It is a shift from individual responsibility to experienced group responsibility for outcomes. To be successful, team members must feel responsible for both their own work and the work of the other team members. Only when team members experience both types of responsibility will they work in a cooperative way.

Task interdependence can come from the distribution of skills among team members and the work processes of the team. It is one reason why action teams (e.g., sports teams) and cross-functional teams (e.g., design teams where members have different skills) often are more successful than student project teams. In a sports team, the players need one another to succeed. In a cross-functional team, working together is the only way to complete a project. However, in a student team, the students typically all have the same skills and knowledge, and they do not need one another to complete the task.

Interdependence helps motivate team members in several ways. When team members depend on one another to complete a task, power is shared among the members (Franz, 1998). The more team members need one another to complete a task, the more power each team member has over the group. Task interdependence affects how factors such as conflict, cohesiveness, work norms, and autonomy relate to group effectiveness (Langfred, 2000). When teams are highly interdependent, these variables have a more powerful effect on how well teams perform. Interdependence also encourages members to believe that their contributions to the group are indispensable, unique, and valuable, thereby making them more willing to put effort into the group's task (Kerr & Bruun, 1983).

Evaluations and Rewards. Interdependence relates to both the task and the outcome of the team's work. The task may require coordinated effort, but team members may believe their evaluations and rewards are primarily based on individual performance, rather than on the success of the team's effort. Research shows that a belief in outcome interdependence is important because it helps motivate members to work together (Van der Vegt et al., 1998).

To be successful, team members must feel responsible for both their own work and the work of other team members. Group goals and group reward systems encourage this dual sense of responsibility. For example, managerial teams often do not perform well because managers are more concerned about what happens in their respective departments than in the organization as a whole. One of the values of companywide profit-sharing programs is to make organizational success an important goal that is rewarded for each member of the managerial team. This encourages the managers to think about what is good for the organization, rather than only about what is good for their departments.

A balance of individual- and team-based rewards is necessary to encourage both a commitment to the team and an incentive for individual performance (Thompson, 2004). Finding the right balance can be difficult for an organization. In addition, the performance evaluation system must fairly identify both team success and an individual's contribution to that success. When individual contributions to the team are identifiable and linked to the reward system, motivation is increased (Harkins & Jackson, 1985). (The topic of evaluating and rewarding teams is discussed in more detail in Chapter 18.)

Goals. As discussed in Chapter 3, team goals help provide direction and motivation for the team. Clear goals support motivation by leading to increased effort, better planning, better performance monitoring, and increased commitment to the group (Weldon, Jehn, & Pradhan, 1991).

Goals affect team members' sense of efficacy, which is the belief that one has the ability to successfully complete a task (Bandura, 2000). Self-efficacy often is reduced in a group because members do not believe their contributions are important or valued. This reduces motivation toward the team's task. Increasing a sense of team or collective efficacy, which is the belief that the team is able to complete its task, helps increase motivation. Teams with higher collective efficacy have higher levels of motivation to perform, greater staying power when they encounter difficulties and setbacks, and improved performance.

Team efficacy has a reciprocal relationship with team performance; in other words, successful performance increases team efficacy and vice versa (Ilgen, Hollenbeck, Johnson, & Jundt, 2005). Team efficacy is influenced by a number of factors (Burn, 2004). Teams that have been successful in the past have higher levels of team efficacy. Leaders who believe their team is competent create teams with higher collective efficacy. Teams with higher collective efficacy are more likely to set higher performance goals, which encourages greater performance.

Commitment and Cohesion. The more people value membership in the group, the more motivated they are to perform. The increased sense of commitment and attraction to a group is called group cohesion. Cohesive groups are less likely to experience social loafing (Karau & Williams, 1997). Group cohesiveness includes a commitment to the task that the group is performing. In a highly cohesive group, members like the task the group is performing, enjoy working together on the task, have personal involvement in the task, and take pride in the group's performance. Highly cohesive groups have more commitment to their tasks and perform better (Wech, Mossholder, Steel, & Bennett, 1998). Because group cohesion has important effects other than motivation, a more complete discussion of it is presented in the next section.

2. Group Cohesion

Group cohesion refers to the interpersonal bonds that hold a group together. Cohesion is a multidimensional concept (Beal, Cohen, Burke, & McLendon, 2003). To many theorists, group pride or social identity is the core of group cohesion. Members of a cohesive group have a shared social identity. Membership in the group is personally important, so they define themselves as members of the group (Hogg, 1992). Others view cohesiveness as a type of social attraction (Lott & Lott, 1965). Members of a cohesive group like one another and feel connected because of this relationship. Cohesiveness also can come from the group's task. The joining together to work as a team can create a sense of cohesiveness (Guzzo & Dickson, 1996).

The sense of identification with the group that occurs in cohesive groups has important implications (Hayes, 1997). A group is better able to manage stress and conflict among its members if it has a firm sense of itself as a distinctive group. The creation of a sense of insiders and outsiders to the group causes people to view the members of the team as similar and at the same time different from members of other groups. In work teams, members may have very different skills, professions, and even statuses, but such differences do not prevent development of a cohesive team.

How Cohesion Affects the Group's Performance

Group cohesion affects the group in a number of ways. People who are part of cohesive groups are more satisfied with their jobs than are members of noncohesive groups (Hackman, 1992). Group cohesion also helps reduce stress because members are more supportive of one another. The interpersonal effects of group cohesion are generally positive, but the effects on a group's performance are mixed.

Group cohesion has a generally positive impact on group performance (Mullen & Copper, 1994). This is especially true for smaller groups. This relationship goes in both directions: Cohesion can help to improve performance, and performance can help to improve cohesion. When a group is successful in its task, its level of cohesiveness increases. Cohesion based on commitment to the task has a larger impact on performance than cohesion based on group attraction or social identity. The effects of cohesion are more important when the group's task requires high levels of interaction, coordination, and interdependence (Beal et al., 2003).

Members of a cohesive group are more likely to accept the group's goals, decisions, and norms. The increased interpersonal bonds among group members increase the pressure to conform to group norms. As was seen in the discussion of group norms, norms can either support or hamper group productivity (Sundstrom, McIntyre, Halfhill, & Richards, 2000). Effective work teams have norms that support high-quality performance and a level of group cohesiveness that provides social support to its members. However, cohesive teams that lack good performance norms may be ineffective and highly resistant to change (Nemeth & Staw, 1989).

Cohesiveness affects a group's social interactions, and this can affect performance and decision making. Low levels of group cohesiveness limit a team's ability to work together. Because they know one another better, cohesive teams are better able to communicate and coordinate their actions (Beal et al., 2003). However, high levels of cohesiveness can impair a team's decision-making ability. Sometimes, group members will agree to a decision not because they agree with it, but because they do not want to upset the group's relationships (Janis, 1972).

An important aspect of group cohesion relates to conflict resolution and problem solving. A team with poor social relations will avoid dealing with problems until they disrupt the team's ability to perform the task or threaten its existence as a team. A team with good social relations is better equipped to handle problems when they arise. The team can do this because its more open communication allows team members to manage conflicts constructively. This is one reason it is important to develop group cohesion and good social relations early in the team's existence. Forming good social relations early means better ability to solve problems and manage conflicts throughout the team's work.

Building Group Cohesion

Research on organizations has identified several factors that encourage cohesion in work teams (McKenna, 1994). Team members in a cohesive group tend to have similar attitudes and personal goals. They have spent more time together, and this increases their opportunity to develop common interests and ideas. A team's isolation from others may help produce a sense of being special and different. A smaller team tends to be more cohesive than a larger team. Having strict requirements to join a team increases cohesion. Finally, when incentives are based on group rather than individual performance, the team becomes more cooperative and cohesive.

Several approaches can increase cohesion in a work team (Wech et al., 1998). Training in social interaction skills, such as effective listening and conflict management, can improve communication and cohesion. Training in task skills, such as goal setting and job skills, improves the team's ability to work successfully. Team success, and reward for success, improves cohesion. The team leader can enhance cohesion by promoting more interactions among team members, reducing status differences, ensuring that everyone is aware of one another's contributions, and creating a climate of pride in the team.

3. Team Roles

Roles are one of the basic building blocks of successful team performance. A role is a set of behaviors typical of people in certain social contexts. Roles within a group are similar to roles in a play; they describe what people are supposed to do and how their parts relate to what others in the group are doing. Group members can negotiate the roles they want to play, and they have a certain amount of freedom in the performance of their roles.

A group can deliberately create roles for members to perform. These roles are task-related and allow the group to operate more efficiently. Even without deliberately creating formal roles, group members assume informal roles within the group, which emerge over time as the group interacts. These roles can be task-related (e.g., expert, facilitator) or socially related (e.g., supporter, clown).

The selection or allocation of roles may occur in a variety of ways. The organization, team, or individual may select roles. For example, management in the organization may assign the team leader, the team may elect its leader, or the team may have no official leader (but an informal leader may eventually emerge in the team). Type of role also affects the selection process. The team often selects members to perform skill-based tasks, whereas social roles often emerge by self-selection.

Definitions of roles also vary. The group explicitly defines some role behaviors, whereas the person filling the role defines other behaviors. For example, the recorder is the person who takes notes prior to writing up the minutes, but whether these are funny or detailed depends on the particular recorder fulfilling the role.

Role Problems

The roles people perform in a group may be a cause of stress. Role ambiguity and role conflict cause this stress. Because group roles often emerge without formal definitions, the responsibilities of the roles often are ill defined. A person fulfilling a role may not understand what other group members expect of him or her, thus creating uncertainty in the role-performer and sometimes hostility from the other group members because the role is not being performed as desired.

Group members also occupy several roles at a time, which may involve conflicting demands. Inter-role conflict occurs when a person occupies several roles that are incompatible with one another. For example, when a person is promoted, he or she often experiences conflict between being a manager and being a friend to former coworkers. Conflicts also may occur within a single role (i.e., intra-role conflict). In a task force with team members from different areas of the organization, members may experience a conflict between doing what is good for the team and doing what is good for their organizational areas.

Role ambiguity and conflict have a negative impact on people in an organization. These role problems can create higher levels of stress, decreased satisfaction and morale, and increased job turnover (Kemery, Bedeian, Mossholder, & Touliatos, 1985). Role problems also decrease commitment to the organization and reduce involvement and participation in the group's interactions (Brown, 1996).

For a project team, role problems often appear to worsen near the end of the project. As team members rush to complete their assignments, they become more aware of the different expectations members have about who will do what. These different expectations of the roles members should be performing lead to conflicts when the team is already stressed about the project deadline.

To address role problems, a team may make explicit the important roles in the group. The tasks that the group is performing may be prioritized so that group members can decide what to do when there are conflicts among tasks.

Types of Team Meeting Roles

Team meetings are an example of how roles are useful for a team. Meetings operate more efficiently when major task roles are explicitly defined and team members are assigned to fulfill them (Kayser, 1990). One of the main meeting roles is that of leader or facilitator. The leader is responsible for structuring the team's interactions to ensure that the team completes its goals. The leader manages the structure of the meeting, but not the content. The primary activities of the leader are (a) develop the agenda to help structure team meetings; (b) ensure that information is shared, understood, and processed by the team in a supportive and participative environment; and (c) remove internal problems that hinder the team's operations.

The recorder takes notes on key decisions and task assignments (i.e., who agreed to do what). The minutes focus on the main points rather than capturing the entire discussion. Good documentation of team actions is important; it makes information available for future use, provides a fixed point of reference, and provides evidence of the analysis process used.

Sometimes the recorder acts as the team scribe, noting comments of team members on a blackboard or flip chart during a discussion. After the discussion, the recorder notes the conclusions reached by the team. Some groups assign the role of scribe to another member or to the leader.

Another role is team timekeeper. When the agenda is presented at the beginning of the meeting, teams may identify the time allotted for each item. The timekeeper reminds the team when it has used the time allotted for an agenda item. The team may continue on that topic, recognizing that to do so will extend the meeting longer than planned.

These team meeting roles should be recognized and filled at the outset of a team's existence. However, team members need not be permanently assigned to these roles initially. It is often better to rotate people through roles, giving everyone a chance to try them out, before the team assigns permanent roles (Kayser, 1990). There are several benefits to this rotation. Every team member has the chance to practice roles. It is a good learning experience, enabling members to fulfill other roles later if there are team absences. In addition, the team has a chance to see how everyone performs. After team members have tried out roles for several meetings, the team is better able to select who should fill them on a more permanent basis.

4. Task and Social Behaviors

Groups perform two basic types of behaviors: task behaviors and social behaviors. Task behaviors focus on the group's goals and tasks. Social behaviors focus on the social and emotional needs of the group members. They help maintain social relations among the members and sometimes are called "group maintenance behaviors." To function effectively, groups need both task and social behaviors. Table 4.1 shows the primary task and social behaviors that occur during group interactions (Benne & Sheats, 1948).

The optimum balance between task and social behaviors depends on the characteristics of the group (Benne & Sheats, 1948). In task-oriented groups such as work teams, task-oriented behaviors will dominate the group's interactions. A study of engineering teams found that more than 90% of a group's interactions were task-oriented (Levi & Cadiz, 1998). When technical teams are under time pressure, they may not have time to devote to

TABLE 4.1

Types of Group Behaviors

Behavior	Function
Task Behaviors	
Initiator/ contributor	Proposes new ideas or new ways for the group to act
Information giver	Provides data and facts for decision making
Information seeker	Requests more information to help in making decisions
Opinion giver	Provides opinions, values, and feelings
Opinion seeker	Requests the opinions of others in making decisions
Coordinator	Shows relationships of ideas to organize the discussion
Energizer	Stimulates the group to continue working
Evaluator/critic	Questions the group's ideas and procedures
Social Behaviors	
Encourager	Supports and rewards others
Harmonizer	Mediates conflicts among members
Compromiser	Shifts their position in order to reduce conflict
Expediter	Facilitates communications from others
Standard setter	Evaluates the quality of the group's interactions
Follower	Accepts ideas of others
Group process observer	Observes and comments on the group's processes

SOURCE: Benne, K., & Sheats, P. (1948). Functional group members. *Journal of Social Issues, 4*, 41–49.

group process issues. Under these conditions, teams may fall back on traditional management methods rather than using teamwork to get the job done (Janz, Colquitt, & Noe, 1997). In general, effective work teams spend about 80% of their time working on the task (Wheelan, 2005).

The right mix of task and social behaviors also depends on the maturity level of the group. When groups are in the forming stage, they must engage in more social-oriented behaviors to develop the social relations of the group. Groups in the performing stage will be dominated by task-oriented behaviors. When a work team develops good social relations early in a project, the team is better able to handle the time pressure at the end because it has developed the working relationships it needs to complete the project.

Value of Social Behaviors

A team tends to focus on the task and ignore the social or relationship aspects of teamwork. Not only does the team fail to promote social relations, but many team members do not even believe they are necessary. It is important to recognize that a team needs a balance. Social behaviors are important for building trust in communication, encouraging the team to operate smoothly, providing social support, and rewarding participation. When a team runs into problems, it often blames individual team members and does not recognize that weak social relations in the team may have caused these problems.

Although team members and managers often state that task skills are more important than social skills or likeability, they do not select new team members on that criterion. Teams more often select likeable people with limited skills than they do competent people who are difficult to work with (Casciaro & Lobo, 2005). Social relations are vital in work teams. Members who like their teammates will try to get the best possible performance out of those individuals. On the other hand, it can be so difficult or unpleasant to get information and assistance from people who are difficult that members avoid asking them to participate.

There is no formula for the right balance of task and social behaviors. Some teams operate well when most of their behaviors are task-oriented. A team is out of balance when emotions or personality conflicts become disruptive to team operations. Such behaviors indicate a breakdown in social relations.

Observation studies on task teams show that one deficit of team communication is lack of praise, support, and positive feedback (Levi & Cadiz, 1998). All team members are responsible for this lack of positive communication. Members are quick to criticize another member's idea if they do not like it, but they are reluctant to praise a team member for a good idea or even for good performance. Increasing positive support by team members greatly helps improve social relations within the team and increases its effectiveness.

One value of using a team to perform a task is that groups reduce stress by providing social support to its members. The types of social support provided by groups are shown in Table 4.2. Stress may disrupt a team's ability to perform and may encourage members to leave the team.

Improving Team Interactions

Group process observation and analysis may improve a team's interactions. The group process observer provides valuable support by observing and commenting on how the group is operating. Many team-building

TABLE 4.2

Types of Social Support Provided by Groups

Emotional support	Rewarding and encouraging others
	Listening to problems and sharing feelings
Informational support	Giving ideas, advice, and suggestions
	Explaining and demonstrating how to perform a task
Task support	Helping another with work tasks
	Providing supportive actions
Belonging	Expressing acceptance and approval
	Demonstrating belonging to the group

SOURCE: From *Group Dynamics*, 3rd edition by Forsyth, D., copyright © 1999. Reprinted with permission of Wadsworth, a division of Thomson Learning, www .thomsonrights.com. Fax 800 730–2215.

programs use outside group-process observers to evaluate group interactions and advise the group on improving its performance. Although this is a valuable function, it is better if the members of the group do it themselves (Dyer, 1995). Developing group-process observation skills among members allows the group to work on its problems when they occur rather than waiting for an outside consultant.

When a team analyzes its group process, a number of common problems emerge (Hayes, 1997). In most cases, the team uses only a limited range of available behaviors. For example, team members might frequently give opinions, but only rarely provide support for the ideas of others. Team members also can become stuck in behavioral patterns rather than responding to the needs of the team. For example, one member may become the team's critic and rarely provide information to foster decision making. The team's performance improves when people are more flexible, using behaviors more suited to team needs than to their personal behavioral styles. The use of group-process observations may help group members see what is lacking in their interactions, and may encourage team members to adjust their personal styles to enable the team to operate more effectively.

Group process analysis has been used to develop team-building programs and analyze the characteristics of effective teams (Belbin, 1981). An effective team requires a balance of task and social behaviors, and the right balance depends on the nature of the task. A team must learn how to shift its emphasis depending on its current needs. Because the personalities of team members will limit their range of their behaviors, the team needs people with different interpersonal skills to ensure a balance of behaviors. The team cannot effectively use its technical expertise without a balance of task and social behaviors.

Summary

Motivation is a problem for many groups. Working in a group can encourage social loafing, which is the reduction in individual effort that occurs when the individual is performing in a group. Free riders and the sucker effect are related motivational problems. These motivation problems may be caused by tasks that do not require coordinated efforts, inability to identify individual contributions to the group's work, and the false belief that individual members are doing their fair share.

Improving group motivation requires countering the negative effects of social loafing. The group's task should be involving and challenging and should require coordinated effort to be completed. The group evaluation and reward system must recognize and reward both individual and group performance. The group's goals should create the belief that motivated effort will lead to success. Finally, strengthening commitment to the group by increasing cohesion helps to increase group motivation.

Group cohesion is the interpersonal bond that forms within a group. It can emerge from feelings of belonging, social identification, interpersonal attraction, or commitment to the group's task. In most cases, a cohesive group performs better than a noncohesive group because of improved coordination and mutual support. However, high levels of group cohesion may sometimes encourage conformity and impair decision making. One of the main ways of developing group cohesion is to improve communication within the group.

Roles are sets of behaviors that people perform in groups. They may be deliberately created and filled, or may operate on a more informal basis. Ill-defined roles (i.e., role ambiguity) and conflicts among roles may create stress for group members. Formal team roles (e.g., leader, recorder, timekeeper) help a team to operate more efficiently.

Group members perform task behaviors and social behaviors. Task behaviors help the group perform its task, whereas social behaviors maintain the group's interpersonal relationships. Work teams often ignore the importance of social behaviors, leading to reduction in interpersonal support and an increase in stress. Team interactions may be improved by better balancing the types of behaviors performed. Also, team members must learn how to act as process observers to improve the team's interactions.

TEAM LEADER'S CHALLENGE—4

You are the student team leader in a senior-level engineering lab course. The team has seven members and is halfway through a semester-long project that is not going very well. Although some students are highly motivated, a few slackers are creating a discouraging atmosphere for the rest of the team. At the last team meeting, a project discussion turned into a heated, and personal, argument. Since then, relations among several students have been strained.

One aspect of the argument among the students was about who is responsible for various tasks. Multiple team members are addressing some tasks, while other tasks are being neglected entirely. As the team leader, you are uncertain who is responsible for these missing assignments. You need to intervene to get the team back on track.

What team problem should you focus on first for improvement?

How should you try to improve the performance of the team?

Is the primary problem social or task related?

ACTIVITY: OBSERVING TASK AND SOCIAL BEHAVIORS

Objective. A team needs both task and social behaviors to operate effectively. Task behaviors help the team complete its goals. Social behaviors foster communication and maintain team social relations. Team members often vary in how much they participate and in the types of behaviors they perform.

Activity. Observe the communication in an existing team or a group discussion of the Team Leader Challenge. Using Activity Worksheet 4.1, note whether each communication is task-oriented or social-oriented. Record each communication by noting how frequently task communications and social communications are contributed by each team member.

Analysis. How does the frequency of task communication compare with the frequency of social communication? Was there a balance between these types of behaviors? How is the use of task communication and social communication distributed among team members? Are there people who are primarily task-oriented or social-oriented? Is the leader primarily task-oriented or social-oriented?

Discussion. What is the right balance between task behaviors and social behaviors in a team? What factors cause this balance to change? What type of behaviors should the leader perform?

ACTIVITY WORKSHEET 4.1

Observing Task and Social Behaviors

	Team Members					
Task Behaviors	*1*	*2*	*3*	*4*	*5*	*6*
Gives suggestions, opinions, or information; or asks questions.						
Organizes the discussion, or helps the decision process.						
Social Behaviors						
Shows support, satisfaction, or acceptance.						
Encourages communications from others and tries to reduce tension.						

5

Cooperation and Competition

Cooperation is necessary for teams to operate smoothly and effectively, and a cooperative atmosphere offers many benefits for team members. However, many team members find themselves in mixed-motive situations that include both cooperation and competition. Team members may be competitive for cultural, personal, and organizational reasons. Cooperation can be encouraged through strategies focused on group goals, communication, and interpersonal actions. However, if teams become too cooperative, overconformity and poor decision making can result. Competition can create negative effects on the team, even when the team is successful.

Learning Objectives

1. What is the impact of being in a mixed-motive situation?

2. Why do people act competitively in teams?

3. How are cooperators, competitors, and individualists different?

4. How does competition hurt a team?

5. How does competition between groups affect a team?

6. What are the benefits and problems of cooperation?

7. How do teams respond to competitive versus cooperative rewards?

8. How can a team deal with the negative effects of competition?

1. Teamwork as a Mixed-Motive Situation

The essence of teamwork is the cooperative interactions of team members. Cooperation is limited by competition, especially when the goals are not shared. Team members should be working together toward a common goal, but competition makes team members work against one another when their individual goals become more important than the team goal. In a competitive relationship, the goal is to do better than others. When this occurs in a team, the team is prevented from focusing on its common goals.

Being a team member should encourage people to act cooperatively, but team members often find themselves in a mixed-motive situation. Consider the following examples:

You are the member of a budget committee which must allocate funds to various departments within an organization. As a committee member, you want to do what is best for the organization, but you also want to make sure that your department gets more than its fair share of funds.

As a student working on a group project, you want to do a good job so that you can get a good grade. However, you have other classes and demands on your time. What you really want is to put in the least amount of effort and still get a good grade.

As a basketball player, only the team's score determines the winner. You should be focused on coordinating your plays with the other team members. However, there is a scout in the audience, and being the game's high scorer will get you the attention you want.

These are examples of the all too common mixed-motive situations in which team members find themselves. Rather than being cooperative or competitive situations, they are both simultaneously. They create "social dilemmas" for the participants. Each member wants to maximize his or her rewards and minimize his or her costs. Selfish behavior may be the best strategy for each individual, but the team and everyone in it would be better off if people acted cooperatively instead.

Unfortunately, many people decide to be competitive in a mixed-motive situation. Once they start acting competitively (or putting in reduced efforts for the group), others respond in the same way. The end result is poor group performance. This is one of the reasons why students complain that the worst problem with group projects is that not everyone does his or her fair share (Wall & Nolan, 1987).

Cooperation in a mixed-motive situation is encouraged by several factors. When team members believe their contributions to the team are valuable and important, they are more likely to contribute (Kerr & Bruun, 1983). Members are more likely to act cooperatively if they believe others are likely to cooperate (Dawes, 1988). Smaller groups are more likely to be cooperative than are larger groups (Kerr & Bruun, 1983). Finally, the more members trust one another and believe that others will work for the group, the more committed they become (Parks, 1994).

2. Why Are People in Teams Competitive?

Even though working cooperatively on a team should prevent competition, competition may occur anyway. Team members may misperceive the situation and turn a cooperative situation into a competitive one. People sometimes choose to act competitively even when it is in their best interests to act cooperatively. Why do people misperceive a cooperative situation and turn it into a competitive one? The explanations for this phenomenon have to do with culture, personality, and organizational rewards.

Culture

One way to view cultural differences is along an individualist-collectivist dimension (Hofstede, 1980). Individualists tend to be more competitive with their coworkers. The United States has an individualist culture that promotes competition. Our emphasis on individualism, freedom, capitalism, and personal success all support the value of competition. Although we are not "anti-cooperation," we glorify the winners in a competition. To some Americans, saying that competition is bad is un-American.

Clearly, this cultural value affects the ways in which people respond to situations. Some Americans even have a negative attitude toward teamwork because they believe the individual is more important than the group. To them, a focus on the group means loss of individual freedom and autonomy.

From a cultural and business perspective, the Japanese have developed a good approach that combines cooperation and capitalism (Slem, Levi, & Young, 1995). Their collectivist culture encourages cooperation; cooperation is highly encouraged and rewarded, and commitment and loyalty are the keys to success in Japanese corporations. At the same time, Japanese business workers have a keen competitive sense. They believe that they are in a competitive fight for survival with other organizations. The key to this struggle is to band together to overcome these external forces.

Personality

Some people are more competitive than others and act more competitively regardless of the situation. They misperceive situations and redefine them as opportunities to act competitively. This individual difference can be explained as a personality difference. Researchers have identified three personality types that can be used to explain why some people are competitive (Knight & Dubro, 1984). These personality types affect how people interpret the situations they are in and how they define success. Figure 5.1 shows how these personality types relate to individual concerns.

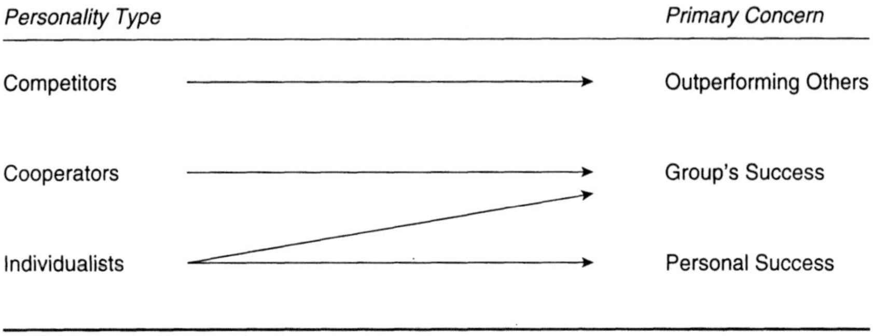

Personality Type		Primary Concern
Competitors	⟶	Outperforming Others
Cooperators	⟶	Group's Success
Individualists	⟶	Personal Success

Figure 5.1   Personality Type and Competition

Cooperators focus on the group. They are concerned with both their own outcomes and those of others. They attempt to make sure the group is successful and that rewards are distributed equitably among group members.

Competitors view a situation as an opportunity to win. They define success not in terms of their individual goals or the group's goals, but rather relative to others' performance. To a competitor, success means performing better than others. Whether they succeed or the group succeeds is less important than whether they do better than the other members of the group.

Individualists define success relative to their own personal goals. Unlike competitors, they do not evaluate their performance relative to others. They may or may not care about the success of the group. The group's success is important only if they have adopted the group's goals for themselves.

Rewards

In many organizations, the shift to teamwork has been disrupted by its human resources practices. Although managers say they want all employees to work as a team, organizational practices often do not encourage or reward teamwork. In most organizations, performance evaluations are based on individual performance, and evaluation is relative to the performance of the other employees. Employees receive a mixed message: Do what the manager says is important (teamwork) or do what you will be rewarded for (standing out as superior to your coworkers). It does not take a psychologist to determine how most employees will respond to this mixed message.

The inability to share rewards is probably the biggest problem that encourages unhealthy competition within organizations (Hayes, 1997). It affects both individual employees and organizational units. For example, departments within an organization often act competitively because they believe they must fight for their share of the organization's resources. Working interdependently to succeed should encourage cooperation over competition (Cheng, 1983). However, many employees are concerned that what is good for the organization overall might not be best for them.

Do managers actually model teamwork to their subordinates? Often managers are the most competitive employees, acting in highly political ways in order to succeed (Gandz & Murray, 1980). This demonstrates to everyone in the organization that competition is key to success. With this kind of message, it is difficult to argue that employees should act cooperatively.

Although culture, personality, and organizational rewards all encourage competition when it is least appropriate, the most important of these factors is the organizational reward system. One may blame American culture, but there are many successful team-oriented organizations in the United States. Obviously it is possible to have a corporate culture that supports teamwork. By definition, personality differences are stable and enduring traits that are difficult to change. Explaining competition as a personality trait implies there is nothing the organization or team can do to change this. However, organizational and team rewards can be changed. Certain approaches to performance evaluation and reward encourage teamwork and discourage individual competition. (Evaluation and reward approaches are discussed in Chapter 18.) Because evaluations and rewards can be changed by an organization, it is the most useful explanation to consider.

3. Problems With Competition

What is wrong with competition? Why should competition not help to motivate a team? Problems with competition occur on both individual and group levels. Individual competition disrupts the group's focus on its common goals. Group competition also creates problems regardless of the group's success. Understanding the dynamics of competition can help explain when and where competition is appropriate.

Communication and Goal Confusion

When individuals or teams in an organization compete against each other, changes occur that prevent the team from being successful (Tjosvold, 1995). Individual competition creates confusion about goals. Eventually, this creates distrust that reduces communication within the team.

A successful team has members who work together to reach a common goal. This common goal provides a focus for the team. However, when team members compete against one another, individual goals can conflict with the team goal. Conflict exists between doing what is best for the individual to succeed (by being better than the others) and doing what is best for the team. This goal conflict creates confusion about the goals of the team. Team members then distrust one another because they are uncertain of one another's motives.

The distrust created by mixed goals leads to reduced communication within a team. Communication requires trust; without trust, there is no reason to communicate with others. Over time, internal competition reduces communication within the team.

This goal confusion and breakdown in communication caused by competition can be seen at the organizational level. The managers in an organization must get together and decide on budget allocations. Should a department manager try to do what is best for his or her department or what is best for the organization as a whole? If the departments are competing for limited resources, should the managers request what their departments need, or should they assume that all the other departments are trying to get ahead and aggressively bargain to get as much as they can? Is there any reason to trust the budget estimates from other departments? This question leads to budget battles where false numbers are used to justify competitive positions.

Intergroup Competition

Intergroup competition can be as much a problem for a team as individual competition. The classic research project on intergroup competition is Sherif's (1966) studies of boys at summer camp. Researchers divided the boys attending the camp into two groups. For a few weeks, these groups of boys competed against each other in a variety of activities. The effects of competition were negative for both groups. The boys who were arbitrarily divided began to see the members of the other group in negative terms. They formed prejudices, that is, negative beliefs about the abilities of the other group and the personalities of its members. Conflicts became a regular occurrence and required intervention by camp counselors.

The problem of competition leading to conflict and hostility is more pronounced in the intergroup situation. Groups are more likely to act competitively with each other than are individuals (Insko et al., 1994). One explanation for this comes from social identity theory (Tajfel & Turner, 1986), according to which a person's sense of self-worth is connected to the groups to which he or she belongs. Consequently, it becomes necessary to view the group as adequate and superior. This leads to an in-group bias, in which group members view their own group in overly positive terms and out-groups in overly negative terms. When the superiority of the group is challenged, members rally to support it and attack the out-group. The conflict escalates easily because group behavior is more anonymous, with fewer interpersonal connections between members and the out-group.

Sherif's (1966) classic study demonstrates several important points about the effects of competition. The main focus was to show that competition led to prejudice. However, the study also revealed the effects of external competition on a team. When a team enters a competition, the team experiences an increase in cohesion and group spirit. Team members become more task-focused and tolerate more autocratic leadership. As the competition continues, more loyalty and conformity are demanded from team members. In the short run, these changes may increase productivity and efficiency. In the long run, however, problems arise for the team, regardless of its success.

A team in a competition focuses on its task to the exclusion of social and emotional issues. Over time, ignoring these social issues can lead to the breakdown of the team. Demanding loyalty and conformity from team members may hurt the team's ability to adapt to change. Creativity and innovation may be stifled by competition.

These negative effects of competition occur for both winners and losers. When groups compete, the winners attribute their success to their own superiority (Forsyth & Kelley, 1996). This causes the winners to ignore their problems, which go unsolved. The losing teams often enter into a period of blaming and scapegoating (Worchel, Andreoli, & Folger, 1977). They first blame their losses on the situation, then team members blame one another. Eventually, if they survive the internal emotional turmoil, the losing teams are able to recognize and solve their problems.

One of the amazing things about the Sherif (1966) study is how easy it is to replicate. In a couple of hours in a workshop, a group of people can be divided into competitive teams. All the negative emotional effects of competition (e.g., prejudices, conflicts, and misperceptions that disrupt performance) soon emerge. This occurs regardless of whether one is studying boys at camp or executives in training programs (Blake & Mouton, 1969). People fall easily into these negative behavior patterns, and even accept the negative effects of a competition if they win.

When Is Competition Appropriate?

Competition is the basis of capitalism because competition encourages innovation, lower prices, and motivation. Given the negative effects of competition, how can this be a true statement? It is important to recognize the difference between internal and external competition. Capitalism is based on competition between organizations, not within organizations. It is useful that Ford competes against General Motors in producing high-quality, low-cost automobiles. It is not useful for Ford's accounting and manufacturing departments to compete for status within the organization.

To understand how competition affects teamwork, it is important to make this distinction between what occurs within and outside the team. Competition between organizations can help improve productivity (Hayes, 1997). Competitors provide motivating goals and feedback about performance. However, competition inside an organization can be devastating. It does not matter if the marketing department at General Motors does not give accurate information to Ford; Ford does not expect it. But when departments within an organization lie to each other to get ahead, the negative impact can be substantial.

Competition may be positive within an organization when jobs are independent rather than interdependent. For example, an organization may sponsor competitions among its sales staff and reward the best performers. This works when salespeople do not depend on one another to make sales (and when rules are in place that prevent the sabotaging of sales by others). However, most jobs within an organization are interdependent. This is especially true when the organization uses teams. Internal competition among teams within an organization can lead to sabotaged work, unjustified criticism, and withholding of information and resources (Tjosvold, 1995).

4. Benefits of and Problems With Cooperation

Cooperation offers many benefits to both the team and its members. However, in some situations, too much cooperation can disrupt a team's performance and decision-making abilities. It is necessary to understand how teams respond to the use of competitive and cooperative rewards.

Benefits of Cooperation

Competition is good for the winners. In other words, for the majority of people in a competitive situation, competition is not a good thing. When members of a group compete, the winners are motivated by the competition.

Some members who believe that they have a chance of winning are also motivated. However, over time, most of group members (about 90%) stop believing they will win. Therefore, they stop being motivated by competition.

Cooperation has the opposite effect on team members. In a cooperative team, all team members are motivated by the team's goals. This motivation is mutually reinforced or encouraged. Team members help and learn from one another. Not only does the team perform better, but most of the individual members also perform better.

Research on cooperative education demonstrates the benefits of cooperation and competition (Johnson, Maruyama, Johnson, Nelson, & Skon, 1981; Slavin, 1985). In cooperative education classrooms, the best performers still perform at a high level, but the performance of average and lower performers improves. The high performers spend time helping others, and they learn from this experience. Overall, the groups have higher performance, better social relations, higher self-esteem, and a better attitude toward school.

The benefits that accrue to individuals in a cooperative situation have a positive impact on teamwork. Cooperation encourages supportive, rather than defensive, communication (Lumsden & Lumsden, 1997). Team members are more willing to talk to one another, and this encourages more communication. Increased communication improves coordination on tasks, satisfaction with working together, and overall team performance (Cohen & Bailey, 1997).

The benefits of cooperation at work depend in part on the assigned tasks. Cooperation is more important when tasks are ambiguous, complex, or changing (Tjosvold, 1995). Such tasks require substantial information sharing to determine the best way to perform them. Because they require coordination, cooperation also is more important when tasks are interdependent. Most professional and managerial work meets these task requirements.

Cooperation provides the foundation for the social relations of group members. Groups that work cooperatively have less tension, fewer conflicts, and fewer verbal confrontations (Tjosvold, 1995). They also enjoy a stronger sense of team spirit and greater group cohesion.

Problems With Cooperation

Cooperation has its own problems. A team can be too cooperative. It can become so focused on maintaining its internal social relations that it loses sight of the team's goals. Problems with cooperation affect both performance (conformity) and decision making (unhealthy agreement).

Conformity. Highly cooperative groups tend to become highly cohesive. Over time, team members become socially and emotionally connected to one another, which improves communication and coordination. However, it can create problems because the team becomes too oriented toward itself.

A highly cohesive team is self-rewarding. It rewards contributions and discourages behavior that is not accepted by the team. This means that the team demands conformity from its members. Conformity can help the team operate, but it also can make the team resistant to outside influence and changes to the way it operates (Nemeth & Staw, 1989).

When a group is functioning well and has good performance norms, conformity is a benefit. However, conformity can make it difficult to influence the team and change its direction. Even a highly cohesive and cooperative team can perform poorly. Sometimes a work group has norms about not doing too much work (e.g., some of the work groups in the Hawthorne studies noted in Chapters 1 and 3). These norms are enforced by the team and can be resistant to change from the organization.

Unhealthy Agreement. Another negative impact of cooperation involves a team's ability to make decisions. Decision making should be focused on making the best decision, given the constraints of the situation. Cooperation can help decision making by establishing trust, which encourages open communication. However, because a cooperative group is cohesive, members' liking for one another can disrupt the decision-making process.

The Abilene paradox describes a problem with group decision making caused by trying to be friendly and cooperative (Harvey, 1988). This occurs when group members adopt a position because they believe other members want them to. The members fail to challenge one another because they want to avoid conflict or achieve consensus. In the end, they support a proposal no one really wants because of their inability to manage agreement. For example, a project team may continue working on a design strategy that no one thinks will work. However, everyone believes the other team members support this approach, so no one raises objections during team meetings.

The Abilene paradox is an example of unhealthy agreement within a group; the team's desire to reach agreement on an issue becomes more important than its motivation to find a good solution (Dyer, 1995). Team members look for the first acceptable solution, or just go along with the leader's solution to avoid disagreements and conflict. This search for quick solutions and avoidance of conflict can lead to poor decisions that cause problems and time delays later in the project. The team is suffering from unhealthy agreement.

Following are some symptoms of unhealthy agreement:

- Team members feel angry about the decisions the team is making.
- Team members agree in private that the team is making bad decisions.
- The team is breaking up into subgroups that blame others for the team's problems.
- People fail to speak up in meetings and communicate their real opinions.

Competitive Versus Cooperative Rewards

There are benefits and problems with both competitive and cooperative reward systems (Beersma et al., 2003). Competitive rewards are effective in motivating individual performance, while cooperative rewards promote trust, cohesiveness, and mutual support, which promote team performance. When a task requires coordinated effort, cooperative rewards are more effective than competitive rewards. However, this simplistic view of cooperative and competitive rewards ignores several important factors that influence work.

What is the primary performance goal? Accuracy and speed of performance are separate, unrelated criteria. Most complex tasks in organizations require both speed and accuracy, but the relative importance of these two factors may vary. A manufacturing team may be encouraged to produce as fast as possible, but this is likely to negatively affect the quality of their performance. Emergency medical teams often must work quickly, but they must also be concerned with the accuracy of their performance. Competitive rewards are strong motivators, especially for encouraging speed (Beersma et al., 2003). Cooperative rewards encourage discussion, collaboration, and information sharing, which may improve accuracy but will slow the speed of performance.

One of the arguments against using cooperative rewards is that they may encourage social loafing by the team's poor performers (Beersma et al., 2003). High performers are often internally motivated and knowledgeable about how to perform the task, while poor performers have either motivation or skill problems. One of the purposes of organizing people into teams is to enable team members to share their workload and help one another (Ilgen, Hollenbeck, Johnson, & Jundt, 2005). When a member performs poorly because of skill or ability problems, team members will provide assistance and share their knowledge, especially when there are cooperative rewards. When a low-performing member is engaging in social loafing or has motivation problems, competitive rewards may help motivate him or her. Even with cooperative rewards, team members may not provide assistance to someone whose performance is low because of a lack of motivation.

5. Application: Encouraging Cooperation

The central issue in cooperation is the beliefs that team members hold about team goals and the motives of other members (Tjosvold, 1995). Cooperation is based on mutual goals that encourage trust and the ability to rely on others. This encourages team members to combine and integrate their efforts, thereby promoting successful teamwork. Incompatible goals create suspicion and doubt about other team members, thereby leading to a breakdown in communication. Once team members start to compete against one another, the other team members tend to respond in kind (Youngs, 1986).

Encouraging cooperation within a team requires counteracting the negative effects of competition. Competition can lead to confusion about the goals of team members and a breakdown in communication, and strategies for dealing with these effects should focus on developing common goals and rebuilding trust and communication (Figure 5.2). In addition, once a team has established a competitive relationship, team members must develop a strategy for negotiating cooperation in the future.

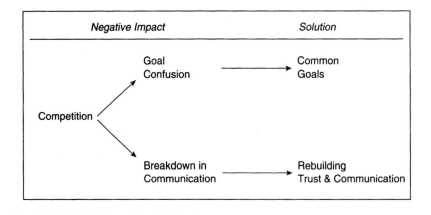

Figure 5.2   Dealing With the Negative Effects of Competition

Common Goals. Research on race relations shows that equal-status interactions can help reduce a sense of competition between groups, but that contact is insufficient by itself (Triandis, 1994). The groups need some reason to work together so as to break down competitive situations. One approach to forming bonds between groups is through the use of superordinate goals (Sherif, 1966). A superordinate goal is a common goal that all of the groups accept as important. By working together on this common goal, prejudice and conflicts between the groups decrease. In the Sherif summer camp studies, counselors brought the boys together to work on common problems that affected the entire camp. Companies focus on the competitive threat from outside as one strategy in encouraging the various parts of their organizations to work together.

Working together cooperatively encourages cooperation in the future. Cooperation encourages a redefinition of the group. Rather than viewing the situation as composed of competing parts, people come to believe that they all are part of the same group or team. However, this works only when the cooperative activity is successful. Failure leads to blaming and scapegoating, thereby furthering the competitive relationship among group members (Worchel et al., 1977).

Rebuilding Trust and Communication. Cooperation is encouraged by trust; competition leads to a breakdown in trust (Smith, Carrol, & Ashford, 1995). Trust has both cognitive (is someone telling the truth?) and emotional (do I feel I can trust this person?) components (McAllister, 1995). Cooperation primarily relates to the emotional component of trust. Honesty, the number of interactions between team members, and the number of helping experiences they have all lead to greater trust among team members. (Approaches to building trust in a group are presented in Chapter 6.)

Cooperation encourages constructive controversy, whereas competition reduces communication and encourages avoidance (Tjosvold, 1995). Constructive controversy allows for open feedback, the raising of questions, and increased communication. Cooperation improves decision making because it increases task-related conflict; members are able to express a conflict openly without creating social problems within the group. By contrast, conflict is avoided in competitive situations because it is too destructive to social relations. Table 5.1 presents a set of communication rules to foster constructive controversy within a team.

Negotiating Cooperation. Research on the negative effects of competition has examined the various strategies that can be used to encourage cooperation with an opponent. There is an entire research field that uses simulation gaming, such as the prisoner's dilemma, to explore the various options. This research has found a fairly simple strategy for encouraging cooperation (Axelrod, 1984).

The problem for people in a competitive situation is how to make the transition to cooperation. Once competition starts within a team, it tends to continue. If a member tries to threaten the competitors, they will become defensive and more hostile. If a member tries to act consistently cooperatively toward the competitors, they will exploit him. An effective strategy must resolve this dilemma.

The most effective strategy has two rules. First, when the opportunity arises, a team member signals his or her desire to form a cooperative relationship by acting cooperatively. The team member should always start by acting cooperatively and creating opportunities to start over again during transition points in the team's existence. Second, the team member should

TABLE 5.1

Rules for Constructive Controversy

1. Establish openness norms.	Encourage all team members to express their opinions and feelings. Do not dismiss ideas because they appear too impractical or undeveloped at first.
2. Assign opposing views.	Assign a person or subgroup the role of critically evaluating the group's current preferences.
3. Follow the golden rule of controversy.	People should discuss issues with others the way they want issues discussed with them. If you want others to listen to you, you should listen to them.
4. Get outside information.	Search for information from a diverse set of outside sources to help the group make a decision.
5. Show personal regard.	Ideas can be criticized, but do not attack a person's motivation or personality.
6. Combine ideas.	The team should avoid either/or thinking, and try to combine ideas to create alternative solutions.

SOURCE: Adapted from Tjosvold, D., Cooperation theory, constructive controversy, and effectiveness: Learning from crisis. In R. Guzzo & E. Salas (Eds.), *Team effectiveness and decision making in organizations.* Copyright © 1995, San Francisco: Jossey-Bass.

always respond in-kind to his or her competitors' moves (i.e., the tit-for-tat rule). If the competitor acts cooperatively, then the member should respond cooperatively. However, if the competitor acts competitively, the member should respond competitively. This is necessary because always acting cooperatively can lead to exploitation.

Summary

Cooperation is the essence of teamwork. However, team members often find themselves in mixed-motive situations that are a combination of cooperation and competition. This is caused by the conflict between individual goals and the team's goals.

People in teams become competitive for three reasons. First, our culture emphasizes the value of competition. Second, people may be competitive, rather than cooperative or individualist, owing to their personality traits. Third, the organization may reward competition among team members. Although all these encourage competition, the organization's reward system is the most important source of competition.

Competition hurts a team by creating goal confusion. Competitive team members focus on individual rather than group goals to guide their behavior. This leads to distrust, which eventually disrupts communication within the team. Competition with other teams also can create problems leading to hostility and conflict. Although competition with outside organizations may be appropriate, internal competition can be destructive to the team and organization.

Cooperation provides benefits for both individuals and the team. Individuals are motivated and supported in cooperative situations. Cooperation encourages communication and interpersonal support in the team. However, cooperation also can cause problems. A cooperative team has high levels of conformity, which can reduce performance and creativity and lead to unhealthy agreement, where team members make bad decisions in order to preserve group harmony. Cooperative rewards are more effective in promoting quality work and encouraging the team to help poor performers.

Several team tactics can be used to deal with the negative effects of competition and to build a cooperative environment. A commitment to common goals helps unite team members. Trust-building activities can be used to help rebuild any breakdown in a team's communication. Finally, certain negotiation tactics can be used to respond to inappropriate competitive behavior in the team.

TEAM LEADER'S CHALLENGE—5

Last year, the manufacturing plant where you work reorganized into self-directed work teams. You are the team leader for one of several teams in the assembly area. The transition to teamwork was difficult, but things have been working fairly well. Several months ago, upper management announced a new team incentive program that rewards each team for exceeding production targets. This incentive program has had mixed effects on team performance and has created some problems at the plant.

Your team is now highly motivated to maximize production, but this has created conflicts with other teams. You find yourself competing with the other manufacturing teams for access to the company's technical support staff. The team has been ignoring machine maintenance issues in order to spend more time producing. There have been several arguments with other teams in the plant who supply parts for your assembly area, and communication with these other teams has deteriorated. You think that the cause of these problems may be the new reward program, but upper management is very supportive of the rewards program.

How should you help the team to deal with their internal problems?
What can be done to improve relations with the other teams at the plant?
How can you explain your team's problems to upper management?

ACTIVITY: COMPETITIVE VERSUS COOPERATIVE GOALS

Objective. Teamwork should be a cooperative activity. Whether team members act cooperatively or competitively depends on how they are evaluated and rewarded. This activity lets groups experience working under competitive and cooperative conditions.

Activity. The class is divided into groups and instructed to make two construction projects. The first project (tower) has a competitive goal, while the second project (arch) has an individualist or cooperative goal. Activity Worksheet 5.1 contains instructions for this activity. It may be useful to assign observers to note how the groups perform these two activities.

ACTIVITY WORKSHEET 5.1
Competitive Versus Cooperative Goals

Construction Projects:

Divide the class into groups of four or more members.

Give each group a box of supplies, including colored paper, magazine, cardboard, tape, scissors, markers, and paper clips.

Project #1—Construct the Tallest Tower

Groups have 15 minutes to construct the tallest, free-standing tower possible.
The group with the tallest tower wins an award (candy?).

Project #2—Construct a Beautiful Arch

Groups have 15 minutes to construct an attractive, free-standing arch.
Groups are allowed to share ideas and materials with one another.
Judges will give awards (candy?) to all beautiful, free-standing arches.

Analysis. How did the groups behave differently during these two activities? What were some examples of cooperative and competitive behaviors that occurred in or between the groups? How did group members feel about participating in the two projects?

Discuss. How do competitive and cooperative goals affect behavior within and between groups? What is the impact of quantity (and/or production) versus quality goals on teamwork?

6

Communication

Central to any team's actions is communication, which can occur in a variety of patterns. The communication process is affected by the characteristics of the sender, the receiver, and the message. The communication climate affects the willingness of team members to participate, so optimizing communication in a team requires managing this climate. Improving communication requires building trust within the team, facilitating team meetings, and developing good communication skills.

Learning Objectives

1. How do the characteristics of the sender and receiver affect a communication?

2. What factors lead to miscommunication in a team?

3. What are the differences between centralized and decentralized communication networks?

4. What are the characteristics of positive and negative communication climates?

5. What biases does a team have when processing information to make a decision?

6. How can one build trust within a team?

7. What are some of the important activities of the facilitator of a meeting?

8. What are the basic communication skills that are useful to facilitate a team meeting?

1. Communication Process

Communication is the process by which a person or group sends some type of information to another person or group. This definition highlights the three basic parts of a communication: sender, receiver, and message.

Sender

The characteristics of the sender or communicator affect the amount of influence a communication will have on the audience. How the audience perceives the communicator affects how the audience interprets the message, how much attention the audience pays to it, and how much impact it will have on the audience's beliefs. The two primary characteristics of the sender are his or her credibility and attractiveness.

The more credible the communicator, the more the audience will believe the message. Credibility relates to the perceived expertise and trustworthiness of the communicator. A credible speaker knows what he or she is talking about and has no motive to deceive. Saying smart things, having credentials, and speaking confidently create perceived expertise. Perceived trustworthiness is affected by the belief that the communicator is not trying to persuade the audience, and will not personally benefit from the communication (Eagly, Wood, & Chaiken, 1978).

Although credibility is a characteristic of the sender, its reception often depends on the audience's evaluation of the speaker. The sender may believe she is an expert without bias, but the receiver might not acknowledge this. For example, a college professor may be considered an expert in his or her field of study, but students may not believe the professor is an expert on various social or political topics. A highly credible communicator may be very persuasive, whereas a communicator with low credibility may cause the audience to believe the opposite of the message (sometimes called the boomerang effect).

The attractiveness of the communicator also influences how a communication is received (Chaiken, 1979). Attractiveness relates not only to physical appearance. We tend to find people more attractive if they are similar to us in appearance, background, attitudes, or lifestyles (Wilder, 1990). We also find high-status people, and others we want to be like, attractive.

Is attractiveness more important than credibility? The answer depends on the issue being examined. If the issue is about objective facts, credibility is more important. If the issue is about subjective values or preferences, identification with the communicator is more important than credibility (Goethals & Nelson, 1973).

Receiver

The receivers, or audience, of a communication vary in ways that affect the influence of the communication. For example, personality characteristics of receivers, such as intelligence, language skills, and self-esteem, affect communication. Intelligence and language skills of the audience determine the wording of the communication. Credible communicators more easily influence those with low self-esteem.

Teams often are composed of members with different professional backgrounds. All professional fields have specialized languages or jargon, thereby making communication across fields difficult. Jargon is inevitable. It helps foster communication among people in a specialty area by allowing them to more easily share complex ideas. However, it can lead to miscommunication and feelings of being left out by recipients who do not know the jargon (Kanter, 1977). This creates communication barriers within teams. To avoid this, language should be clear and simple in mixed audiences.

Receivers also differ in their relationships to messages. It is easy to communicate to someone who already basically agrees with the message. However, if the audience is skeptical or antagonistic, the communicator must be more careful about how he or she frames the message. Low levels of disagreement between the sender and the receiver produce discomfort, and this discomfort often encourages people to change their opinions. High levels of disagreement may cause the receiver to view the sender as not credible and to discount the message. People are more open to arguments that are within their range of acceptability (Zanna, 1993). However, highly credible sources can get away with advocating unpopular positions.

Message

The various characteristics of a message interact with the characteristics of the receiver. Messages can vary in sophistication, level of emotion, and aesthetics. Whether these differences influence the audience depends on how they are perceived.

The basic requirement of a message is that it be understood in order to have an effect. Messages with too much jargon, highly sophisticated language, or complex arguments are ineffective with an audience incapable of appreciating them. However, a communicator may also oversimplify a persuasive communication. One-sided messages that ignore the existence of alternative arguments are no more effective than two-sided messages if the audience will hear other arguments later, or is aware of the other positions (Jones & Brehm, 1970).

Rational arguments have an impact, especially for better-educated audiences (Cacioppo, Petty, & Morris, 1983). Emotionally arousing messages, especially fear-arousing messages, also can be persuasive. However, the communicator must be careful that the audience does not attempt to block or manage the emotion rather than the situation. The sender should provide a way for the audience to act on its emotions so as to work through them.

Communication Within Teams

Characteristics of the sender, receiver, and message are the foundations of successful communication, but they also create the opportunity for miscommunication. The sender may fail to send a message or may not be trusted to send a useful message. The receiver may distort or misperceive the message, or the message may be inaccurate or distorted. There are a number of problems that can disrupt communication within a team.

Senders typically tailor the messages to their audiences, so messages may be shorter or longer depending on assumptions about what the receivers know. For example, when people give directions, they give longer directions to people who are unfamiliar with an area (Krauss & Fussell, 1991). In a team, senders often send briefer messages than are needed because they overestimate receivers' familiarity with the information. Senders are often poor at perspective-taking; they presume receivers have more background information on topics than they really do (Keysar & Henly, 2002). This lack of perspective-taking is one reason why technical professionals such as engineers have difficulty sharing specialized knowledge in a team. They assume that the receivers have sufficient background information to make sense of brief messages.

Messages also become distorted within team discussions. All team members are biased toward presenting information that will be positively received (Higgins, 1999). This causes the team to ignore its problems because unpleasant topics are never addressed. Team members often believe that the reasons behind their statements are obvious, so they do not fully explain issues (Gilovich, Savitsky, & Medvec, 1998). This leads to an illusion that there is clear communication, when in fact there is not.

One of the best examples of clear communication is in military teams. Miscommunication in a military team can be fatal, so the military has developed strict rules about how team members must communicate with and support one another. Successful teams perform the following actions: performance monitoring, feedback, closed-loop communication, and backing-up behaviors (McIntyre & Salas, 1995). Team members monitor the performance of other members and help out when needed. During formal debriefing sessions, all team members give feedback to one another to improve performance. Internal communication is closed-loop; in other words, senders and receivers acknowledge and make sure that the meaning of messages has been correctly received. Finally, team members back one another up; they are trained in each other's jobs so they can replace each other when needed.

6.2 Communication Networks, Climates, and Information Processing

When people talk in a group, patterns of communication develop. In most situations, evenly distributed communication patterns where everyone participates are better for the team. However, patterns created by organizational networks determine where the information flows. Communication climates influence the level and style of communication within a team. Finally, interpersonal processes influence the willingness of team members to share information in team discussions.

Communication Networks

Communication networks are patterns that dictate who may communicate with whom. The nature of these communication linkages has an important impact on group functioning (Shaw, 1978). There are two basic types of networks: decentralized and centralized (Figure 6.1). Decentralized networks include the circle and open models; centralized networks include the Y, wheel, and chain. Centralized networks require any communication to pass through certain members before going to other members. People in decentralized networks have equal access to information, whereas centralized networks create unequal access to information.

Centralized networks provide faster and more accurate information flow on simple tasks; decentralized networks work better on complex tasks (Forsyth, 1999). The problem with centralized networks is due to limits in processing information when flow is restricted. On complex tasks, the key links in the centralized communication networks can suffer from information overload, which reduces their ability to function well.

Most people prefer working in decentralized networks, which have more equal-status interactions. The peripheral members of a centralized network may feel powerless and unappreciated, so they become dissatisfied with the group's communication.

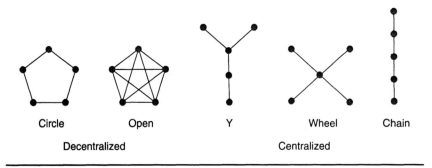

Figure 6.1   Communication Networks

SOURCE: Based on Shaw, M. (1978). Communication networks fourteen years later. In L. Berkowitz (Ed.), *Group processes* (pp. 351–356). New York: Academic Press.

These differences in communication networks are often only temporary because people find ways to work around the networks to complete their tasks and influence the operation of the group (Burgess, 1968). Such informal communication networks operate around the formal networks, thereby mitigating some of their effects.

Communication Climates

Members of effective teams have strong feelings of inclusion, commitment, pride, and trust in their teams. These feelings are developed by a communication climate that is open, supportive, inclusive, and rewarding (Gibb, 1961). Supportive climates encourage people to focus on the message; they allow diverse ideas and expressions of both agreement and disagreement. Because of their emotional comfort, team members are better able to focus on the task. Negative climates are characterized by defensive behaviors. They are closed, alienating, blaming, discouraging, and punishing.

The communication climate develops in cycles (Lumsden & Lumsden, 1997). When team members take a chance and communicate, they can receive supportive responses. These responses encourage trust and openness, which increase their willingness to communicate again, further increasing trust and creating a supportive climate. Communication that receives defensive responses, such as negative evaluations and sarcastic comments, can lead to less trust and greater self-protection. A defensive climate results, which in turn leads to conflict or withdrawal.

Table 6.1 shows the types of communication behaviors that occur in supportive and defensive climates. In supportive climates, messages are facts or opinions rather than negative evaluations or criticisms of others. The focus is on problem solving and discussion of issues, rather than on controlling the team's communication and winning. Instead of guarded statements that reflect concern about power issues, communication is open and expressive, with a free flow of ideas and opinions. Members test out their ideas with the team instead of closing off discussion by drawing overconfident conclusions.

Both defensiveness and supportiveness can escalate. When a supportive climate escalates, it builds openness, trust, and empathy. When a defensive climate escalates, it leads to conflict or withdrawal.

TABLE 6.1
Supportive Versus Defensive Communication Behaviors

Supportive	Defensive
Description	Evaluation
Problem orientation	Control
Spontaneity	Neutrality
Equality	Superiority
Provisionalism	Certainty

SOURCE: From Steven A. Beebe, John T. Masterson, *Communicating in Small Groups: Principles and Practices, 4e.* Published by Allyn & Bacon, Boston, MA. Copyright © by Pearson Education. Reprinted by permission of the publisher.

Processing Information Within the Team

The use of teams creates the potential to make better decisions because members can pool information from diverse backgrounds and experiences. This benefit of using teams occurs only if members share their unique information with the team. However, studies of team communication show that teams spend most of their time reviewing common information and discussing what everyone already knows, rather than combining the unique knowledge and perspectives of members (Gigone & Hastie, 1997). This is why the information held by most team members before a discussion has more influence on a decision, regardless of whether this information is accurate. The focus on common rather than unique information also explains why teams often overlook technical information; such information is likely to be known to few team members, so the team rarely discusses it.

Biases in the ways a team processes information may prevent the team from making good decisions because important information that one member holds is ignored by the group (Stasser, 1992). Consider the example of a design team with members from engineering, marketing, and finance. They should be sharing their unique perspectives and sources of information to create the best design. The engineer should discuss new technical ideas, the marketing person should present the latest results of marketing surveys, and the finance person should examine new options for reducing costs. However, team discussion is more likely to focus on the team's common knowledge and perspectives than on new or unique information. This reduces the team's creativity and can lead to designs with problems that could have been identified in advance.

Avoiding these problems often requires that the team leader actively facilitate the team's communication (Larson, Foster-Fishman, & Franz, 1998). The leader has the ability to ask questions and emphasize technical information so as to focus the team's attention. During discussions, the team should take a problem-solving approach that focuses on encouraging team members to provide factual evidence, not just opinions on the topic. When it is time to make a decision, each alternative should be analyzed in turn to encourage members to discuss the unique information that relates to only that alternative. When technical information is needed, the leader should ask for the opinions of the technical members of the team. Finally, the leader should build trust so that members feel free to contribute to the discussion.

3. Building Trust

The key to good communication in a team is trust. For team members to trust, they must believe the team is competent to complete its task (team efficacy) and the team environment is safe for its members (Ilgen et al., 2005). Trust is the expression of confidence in the team relationship; that is, the confidence one has that other team members will honor their commitments (Thompson, 2004). It is built on past experiences, understanding the motives of others, and a willingness to believe in others. Trust within a team encourages communication and cooperation and makes conflicts easier to resolve.

Trust evolves from shared values, attitudes, and emotions (Jones & George, 1998). We tend to trust people who share our values, and people who are trustworthy tend to trust others more. Trust is also based on the attitudes that people form about one another. For example, are people who work in the organization generally trustworthy or honest? Finally, trust is affected by emotions. Often the decision to trust someone is based primarily on feelings, rather than on concrete behaviors. When trust is broken, it is hard to regain for emotional reasons.

Trust is based on social relationships (Uzzi, 1997). People make investments in developing and maintaining their relationships, and these ties among people encourage cooperation and trust. At the beginning of a social encounter, people take a chance on trusting the other person, while observing how the other responds. The experience of future trust is determined by what happens in the relationship. Trust is built over time through social interactions—through the sharing of feelings and thoughts.

Trust has a direct relationship to interpersonal communication, cooperation, and teamwork. However, it also has a number of indirect relationships (Jones & George, 1998). When teams have high levels of trust, several other favorable behaviors occur that support teamwork. People are more willing to help others in a variety of situations. The free exchange of information is encouraged, and there is increased participation in the team's activities. People are more willing to commit to group goals (and to ignore personal goals) when trust is high. Finally, people are more willing to become involved in the team's activities when trust is high.

Building trust in a group requires performing two types of behaviors: being trusting and being trustworthy (Johnson & Johnson, 1997). Being trusting means being willing to be open with information and sharing with others by providing help and resources. Being trustworthy means accepting the contributions of other group members, supporting their actions, and cooperating in assisting them.

Although building trust is a slow process, trust is quickly and easily destroyed, often by a single incident. Reestablishing trust after it has been broken may be difficult. The following are some techniques to help rebuild trust:

- Apologize sincerely for actions that destroyed trust in the group.
- Act trusting and demonstrate your support for others in the group.
- Promote cooperation in the group.
- Review the group's goals and gain commitment to common actions.
- Establish credibility by making sure that actions match words.

4. Facilitating Team Meetings

In every group, communication is uneven and a few people do most of the talking. In a typical four-person group, two people do more than 70% of the talking; in a six-person group, three people do more than 85% of the talking (Shaw, 1981). Participation in a team discussion is related to the team member's status and personality, the leader's behavior, and the communication climate (Ilgen et al., 2005). Facilitators need to manage the group process in order to equalize participation.

Team meetings often have a structure that helps control and facilitate communication. The model presented in Figure 6.2 was developed for professional business meetings (Kayser, 1990). The meeting starts with a review of the agenda and warm-up activities designed to get people talking socially. The body of the meeting focuses on managing the communication process and making the team's decisions. The meeting ends with a summary of decisions and assignments and an evaluation of how well the team is operating.

Although the team leader is the primary facilitator of meetings, all team members have a responsibility to help facilitate team meetings (Kayser, 1990). The following is a description of the five main communication activities of the facilitator.

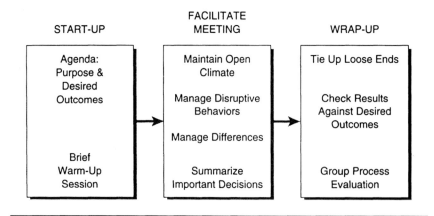

Figure 6.2   Facilitating a Team Meeting

SOURCE: Adapted from Kayser, T. (1990). *Mining group gold.*

Maintain an open and collaborative climate. When discussing the topics of the meeting, the discussion should focus on the issues, not on people's personalities or behaviors. Focusing on individual behavior rather than on issues may make team members defensive and reduce communication. The leader can encourage discussions by asking open-ended questions, organizing discussions around specific issues to focus the team, and asking questions that clarify the situation. The leader should acknowledge when the discussion is going well and verbally praise active team participants. If the discussion gets too personal and individuals are verbally attacked, the leader must intervene and focus the discussion on the issues.

Manage disruptive behaviors. When team members are being disruptive, the leader must be firm but friendly in confrontations. Disruptive team members may dominate the discussion, or be overly talkative or rude to other team members. All team members share responsibility for handling difficult members; it is not just the job of the leader to maintain the flow of the meeting. The leader should acknowledge and verbally reward acceptable behaviors. If problem behaviors persist, the leader should talk privately with repeat offenders. If none of these approaches works, assistance from outside the team (e.g., a manager responsible for supervising the offender) may be required.

Manage differences. Differences can be a constructive force because they can encourage critical thinking, creativity, and healthy debate. Differences are constructive when issues (and not people) are attacked, team spirit is increased, understanding is enhanced, and achievement takes place. Differences also can be a destructive force. They can create winners and losers, group polarization, and unproductive sessions. Differences are destructive when they lead to personal attacks, repeated negative statements, misinterpretations of people's positions, and stubborn resistance to reconsidering positions. Differences can be managed by clarifying the various points of view, defining areas of agreement and disagreement, and taking steps to resolve differences through problem-solving techniques and consensus decision making.

Summarize important decisions. The leader must keep team members focused on the agenda topics. To keep the group process flowing, the leader should stop after major agenda items and summarize the team's conclusions. This allows for a check on whether all team members agree with what has happened at the meeting.

Evaluate the group process. The leader should hold a group process evaluation at the end of each team meeting to discuss how the meeting operated and whether there are areas for improvement. These group process evaluations provide feedback to the team about its performance and help to deal with problems before they get emotionally out of hand.

One reason teams do not evaluate their group processes is that they prefer to ignore their problems. Team members try to avoid conflict and often are unwilling to tell other team members when they are dissatisfied with their performance. The desire to avoid trouble means that problems remain hidden and unresolved until they become so large that they are difficult to manage. The best way to prevent this destructive pattern of behavior is to conduct regular group process evaluations. A technique for conducting group process evaluations is presented in Table 6.2.

TABLE 6.2

Group Process Evaluations

Have team members answer the following questions at the end of a team meeting. Then have the team discuss the results of the evaluation and how it could use this information.

1. How well is the team performing?

    Very Poorly   1   2   3   4   5   6   7   Very Well

2. What is the team doing well?

3. What areas of improvement are needed?

5. Communication Skills for Team Meetings

There are many communication skills that are useful for team members. This section reviews four skills: asking questions, active listening, giving constructive feedback, and managing feelings.

Asking questions. Many types of questions are useful for promoting team discussions (Hackett & Martin, 1993). In general, open-ended questions encourage discussion, whereas closed-ended questions (e.g., yes/no questions) tend to limit discussion. It is better to ask the team to discuss the pros and cons of an idea than to ask team members whether they agree or disagree with it. After someone has answered a question, it is often useful to ask follow-up questions to clarify the issues. When questions are addressed to the leader, they should often be redirected back to the team to promote discussion.

Being asked a direct question by the leader can be a threatening experience that reduces discussion. Leaders should try to ask questions of the entire team whenever possible. After asking a question, the leader should remember to give team members sufficient time to respond. The leader should reward participation by acknowledging responses. If no one responds, the leader should try rewording the question or going around the room and having everyone comment on it. A lack of response may mean that the question has a bias or is putting some team members on the defensive.

Active listening. The goal of active listening is to provide feedback to the sender of a communication to clarify the communication and promote discussion (Johnson & Johnson, 1997). Good listeners communicate their desire to understand the message and improve their understanding.

The largest barrier to effective listening is evaluation. People often spend time evaluating a communication rather than listening to what is actually being said. Constant evaluation makes the sender defensive and decreases further communication.

Active listening is an approach to improving communication. In this approach, the listener paraphrases what he has heard and asks the sender if this is correct. The paraphrasing should convey the listener's understanding of the communication and not be a simple parroting of the message. This sends a message that the listener cares about understanding the message and allows the sender to clarify the communication if needed. Although this is a useful technique, it can become tiresome if used all the time.

Giving constructive feedback. Everyone needs feedback to improve performance. However, receiving feedback (especially negative feedback) may be an uncomfortable experience. Improving one's ability to give constructive feedback is an important teamwork skill (Scholtes, 1994).

The first step in learning to give constructive feedback is recognizing the need for it. Both positive and negative feedback are important. Before giving feedback, the context should be examined to better understand why the behaviors occurred. If a situation is emotional, it is best to wait until things calm down before giving constructive feedback. A team member or leader giving feedback should describe the situation accurately, try not to be judgmental, and speak for him- or herself. When receiving feedback, one should listen carefully, ask questions to better understand, acknowledge reception of the feedback, and take time to sort it out.

If a member is giving solely negative feedback, he or she is not being constructive. Expressing solely negative feedback on the performance or ideas of other team members makes them defensive and discourages communication. It is better to reward the ideas and behaviors one wants than punish the ideas and behaviors one does not want. When giving negative feedback to a team member, corrective alternatives should be offered. Also, negative feedback should be given privately to avoid embarrassing the recipient.

There are several techniques that may enable a team to better accept feedback about its performance:

- Focus on the future. Focusing on the past makes people defensive. Focus the information on how to improve future performance.
- Focus on specific behaviors. Providing general information does not help the team identify the changes needed in its behavior.
- Focus on learning and problem solving. The information provided should help the team improve, not just focus on its deficiencies.

Managing feelings. When emotions become disruptive to the operation of the team, they must be managed effectively (Kayser, 1990). People cannot be prevented from becoming emotional, nor should one want to prevent it. When emotional issues are related to the team's task, the issues should be addressed in the team meeting. Emotional conflicts related to personal issues may need to be handled in private. All team members should learn how to handle emotional interactions in the team. The following is an approach to managing feelings during team meetings.

1. Stay neutral: People have a right to their feelings. The team should encourage and acknowledge the expression of feelings.

2. Understand rather than evaluate feelings: All team members should be sensitive to verbal and nonverbal messages. When dealing with emotional issues, it is best to ask questions and seek information to better understand the feelings.

3. Process feelings in the group: When the team's operation is disrupted by emotions, the team should stop and be silent briefly to cool down. Then the task-related issues should be discussed as a group.

This approach to managing emotions is useful when the emotional issues are related to tasks. Team norms that encourage open communication of emotions increase the benefits to performance of task-related conflict (Jehn, 1995). However, norms that encourage open communication about relationship-oriented conflict have a negative impact on teams. When emotions are about personal or relationship issues, it is not a benefit to process them with the team.

Summary

Communication is one of the central activities of a team. The effectiveness of a communication depends on the characteristics of the sender, the receiver, and the message. Senders are more effective if they are credible and attractive to the audience. Teams often contain members from different backgrounds, so the message must be made understandable to this mixed audience. The basic requirement of a message is that it be stated in such a way that the audience will understand it. Unfortunately, miscommunication is common in teams because teams can lack awareness of others' perspectives.

The communication network, the team's communication climate, and the way teams process information influence team communication. Communication networks define the linkages among team members. Although centralized networks are more efficient for simple problems, most team communication should use decentralized networks. The team's communication climate can be either supportive or defensive, and this has a strong impact on the willingness of team members to participate. Teams should be able to pool the knowledge of each member to make better decisions. However, this often does not occur because of a tendency to focus on common, rather than unique, information.

Trust is a key factor underlying team communication. Trust evolves from the relationships among team members and supports both communication and cooperative behaviors. Developing trust takes time, but destroying trust can happen quickly. Once trust has been broken, it is difficult to rebuild.

Team meetings operate more effectively if a facilitator structures the communication. The role of the facilitator is to maintain an open and collaborative climate, manage disruptive behaviors, manage differences, summarize important decisions, and evaluate the group process.

A number of important communication skills are useful for team members to learn and perform. Asking open-ended, nonthreatening questions fosters better team interactions. Active listening helps clarify the communicator's meaning and acknowledge the importance of the message. Giving constructive feedback is a technique that helps team members learn to improve their performance. Teams can be disrupted by emotions; learning how to process emotions in a group is an important skill.

TEAM LEADER'S CHALLENGE—6

You are the leader of a customer service improvement team that meets weekly at the end of the workday. Early in the team's life, the team had some communications skills training. You closely follow the analysis and decision-making structures from the company's Customer Service Improvement Manual. Over time, as the team has become more comfortable with analyzing quality problems and creating solutions, you have been using less structure in facilitating the team meetings.

However, you have begun to notice problems with the meetings. Not everyone is participating, and the discussions are becoming dominated by several of the older male team members. You've noticed that critical personal remarks have tended to silence some of the women team members. An argument that occurred several meetings ago has caused other team members to stop participating during the meetings. Also, discussions tend to drift off topic and seem like repeats of previous conversations.

What should the leader do to get the team's communications back on track?

What is the best way to handle problem team members during the meetings?

Does the team need more skills training, more communication structure, or outside facilitation?

ACTIVITY: OBSERVING COMMUNICATION PATTERNS IN A TEAM

Objective. Communication within a team often develops into patterns and networks. Team members can either speak to the entire team or speak to individual team members. Observing communication patterns reveals whether the team is working collaboratively, developing subgroups, or forming hierarchies.

Activity. During a team meeting or group discussion of the Team Leader Challenge, note when a team member speaks and to whom. The member can either speak to another individual or to the team as a whole. Using Activity Worksheet 6.1, record the team's communication pattern by drawing arrows connecting the various communicators. Use slash marks on the arrows to note additional communications.

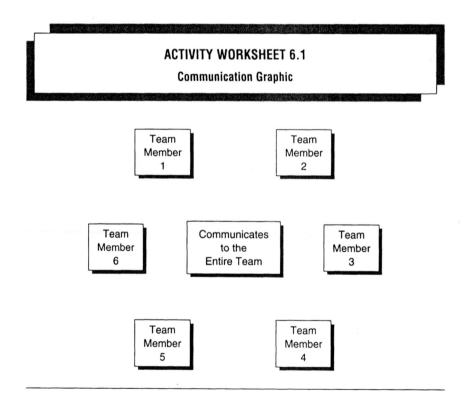

ACTIVITY WORKSHEET 6.1

Communication Graphic

Analysis. Was most of the team's communication to the team as a whole? Did you notice any patterns of communication? Were certain team members more likely to address the team as a whole? Can you determine who the team leader is by observing this communication pattern? How would you describe the communication pattern of the team (i.e., centralized or decentralized)?

Discussion: What should the team leader do to facilitate more equal participation in team discussions?

PART III

Issues Teams Face

7

Managing Conflict

Conflicts of various types are a natural part of the team process. Although we often view conflict as negative, there are benefits to conflict if it is managed appropriately. People handle conflict in their teams in a variety of ways, depending on the importance of their desire to maintain good social relations and develop high-quality solutions.

Managing team conflict can be done through negotiation or with the help of outside assistance in mediation and arbitration. Developing a solution to a conflict requires open communication, respect for the other side, and a creative search for mutually satisfying alternatives.

Learning Objectives

1. Why is the lack of conflict a sign of a problem in a team?

2. What are the healthy and unhealthy sources of conflict?

3. When is conflict good or bad for a team?

4. How does the impact of conflict vary depending on the type of team?

5. What are the different approaches to conflict resolution?

6. Which approach to conflict resolution is best? Why?

7. How is negotiation different from mediation and arbitration?

8. What are the pros and cons of third-party interventions in a conflict?

9. What should a team do to create an integrative solution to a conflict?

111

1. Conflict Is Normal

Conflict is the process by which people or groups perceive that others have taken some action that has a negative effect on their interest. Conflict is a normal part of a team's life. Unfortunately, people have misconceptions about conflict that interfere with how they deal with it. These misconceptions include the following:

- Conflict is bad and should be avoided.
- Team members misunderstanding one another causes conflict.
- All conflicts can be resolved to everyone's satisfaction.

In a dynamic team, conflict is a normal part of the team's activity and is a healthy sign. If a team has no conflict, it might be a sign of a problem. A team without conflict might be suffering from unhealthy agreement, have a domineering leader who suppresses all conflict and debate, or be performing its task in a routine manner and not trying to improve how it works.

Teams often do not handle their conflicts very well. Rather than trying to manage their conflicts, they try to ignore or avoid them. This is called defensive avoidance. To avoid a conflict, everyone becomes quiet when a controversy occurs. Decision-making problems such as the Abilene paradox are in part caused by the desire to avoid controversy. Team members accept what the leader says in order to avoid conflict. The consequences are poor decision making and more problems later in the group's life.

The causes of team conflict change during the team's development (Kivlighan & Jauquet, 1990). During the initial stage, there is little conflict because team members are being polite and trying to understand everyone's positions. This gives way to team conflicts about norms and status issues as the team sorts out its roles and rules. Once the team becomes task oriented, conflicts arise about how tasks should be performed. Often, the final stages of a project have little conflict because team members are focused on implementing the decisions they have made earlier.

It is sometimes better to talk about conflict management than about conflict resolution. Conflict is a normal part of a team's operation, and some conflicts cannot be fully resolved. The resolution of a conflict depends on what type of conflict it is. If it is about task issues, the solution is an agreement. Typically, once the agreement is made, it continues to operate. If the conflict is about relationship issues, then an agreement, periodic checks on how well the agreement is working, and opportunities to redefine the agreement are needed. This is because agreements about relationship issues can change as the relationships change.

2. Sources of Conflict

Conflict may arise from many sources, including confusion about people's positions, personality differences, legitimate differences of opinion, hidden agendas, poor norms, competitive reward systems, and poorly managed meetings. The problem is determining the source to identify whether this is a healthy conflict for the team or a symptom of a hidden problem that needs to be uncovered. If the conflict is about legitimate differences of opinion about the team's task, then it is a healthy conflict. The team needs to acknowledge the source of conflict and work on resolving it. However, sometimes a conflict only appears to be about the team's task and in reality is a symptom of an underlying problem. Finding the root cause of the conflict is important; the team does not want to waste time dealing with only the symptoms of the conflict. Table 7.1 presents a list of healthy and unhealthy sources of conflict.

TABLE 7.1

Sources of Conflict

Healthy	Focused on task issues
	Legitimate differences of opinion about the task
	Differences in values and perspectives
	Different expectations about the impact of decisions
Unhealthy	Competition over power, rewards, and resources
	Conflict between individual and group goals
	Poorly run team meetings
	Personal grudges from the past
	Faulty communications

Legitimate conflicts are caused by a variety of factors. Differences in values and objectives of team members, differing beliefs about the motives and actions of others, and different expectations about the results of decisions can all lead to conflicts about what the team should do. These differences create conflicts, but from these conflicts come better team decisions.

Hidden conflicts that are not really about the team's task may spring from organizational, social, and personal sources. Organizational causes of conflict include competition over scarce resources, ambiguity over responsibilities, status differences among team members, and competitive reward systems. One common type of organizational conflict is the conflict between the team's goals and the goals of individual team members. This is especially true for a cross-functional project team made up of representatives from different parts of an organization (Franz & Jin, 1995). Hidden agendas (i.e., the hidden personal goals of team members) may lead to conflict in the team that can be difficult to identify and resolve. Gaining agreement about the overall goals of the team can help deal with this problem.

Conflict may be due to social factors within the team. A team with a leader who has poor facilitation skills can have poorly run meetings with a lot of conflict. Poor group norms often show up in poorly managed meetings. When meetings are unproductive, conflict may arise because team members are dissatisfied with the team process. Spending time evaluating and developing appropriate norms helps deal with this type of conflict.

Conflicts may arise from personality differences or poor social relations among team members. These may be due to grudges stemming from past losses, faulty attributions, or faulty communication, such as inappropriate criticism or distrust. These are often called "personality differences," but typically their source is interpersonal. Although team members are disagreeing about issues, the root cause of the conflicts is an unwillingness to agree. However, it can be difficult to determine whether someone has a legitimate disagreement about an issue or is opposed to agreeing for personal reasons. To deal with these sources of conflict, team building and other approaches to improving social relations are important.

3. Impact of Conflict

Conflict may have both positive and negative effects on a team. It can help the team operate better by exploring issues more fully, but it can lead to emotional problems that damage communication. Studies on conflict in work teams show that the impact of conflict depends both on the type of conflict and the characteristics of the team (Jehn, 1995).

Benefits of and Problems With Conflict

Although people often view conflict as a negative event, conflict in teams is both inevitable and a sign of health. Teams are organized so as to gain the benefits of multiple perspectives. Team members with these multiple perspectives will view issues differently and learn from one another in the process of resolving their differences. Conflict is an integral part of the team process; it becomes unhealthy for the team when it is avoided or viewed as an opportunity to beat an opponent.

The benefits of conflict are that it encourages the team to explore new approaches, motivates people to understand issues better, and encourages new ideas (Robbins, 1974). Controversies bring out problems that have been ignored, encourage debate, and foster new ideas. When opposing views are brought into the open and discussed, the team makes better decisions and organizational commitment is enhanced (Cosier & Dalton, 1990). When conflict is dealt with constructively, it stimulates greater team creativity. For this to happen, team members must be willing to participate in the conflict resolution process.

Conflict can have negative effects on a team by creating strong negative emotions and stress, interfering with communication and coordination, and diverting attention from task and goals. Conflicts can destroy team cohesion, damage social relations, and create winners and losers who will be a source of conflict in the future. When the conflict is with an outside group, it can encourage a shift to authoritarian leadership, negative stereotyping of others, and an increase in conformity (Fodor, 1976).

Whether conflicts are productive or unproductive depends on how the team tries to solve its conflicts (Witeman, 1991). Productive conflicts are about issues, ideas, and tasks. The team typically tries to solve productive conflicts in a cooperative manner. Unproductive conflicts are about emotions and personalities. The team typically tries to solve these by one side trying to win. In productive conflicts, team members focus cooperatively on solving the issues.

Conflict in Work Teams

Whether conflict has a beneficial or detrimental effect on a work team depends on the type of conflict and the team's task (Jehn, 1995). Both relationship and task conflicts have negative effects on team members' satisfaction with the group. In a team performing a routine task, disagreements about the task hurt group functioning. By contrast, in a team performing a nonroutine task, conflicts about the task can have a positive effect on the team.

Professional project teams are examples of teams performing nonroutine tasks. For this type of decision making or creative team, conflict is a sign that diverse opinions are being presented. The team benefits from this diversity, and conflict helps improve the quality and creativity of decisions. However, when a conflict becomes intense, it can be detrimental to the team. Conflict may reduce the team's ability to reach consensus and may hurt emotional acceptance of the team and its decisions (Amason, 1996).

For a production or service team, the impact of task-related conflict depends on the type of task the team is performing (Cohen & Bailey, 1997). Task conflict disrupts performance on a routine task. However, it can improve performance on a nonroutine task. When the team is performing a routine task, task conflict is a sign that its jobs are poorly defined or that team members are unwilling to cooperate and work together. Conflict among team members in this situation usually is not productive. However, when the team is performing a nonroutine task, such as evaluating how to improve quality, conflict is a natural part of the problem-solving process.

Relationship conflict is detrimental regardless of the type of task a team is performing (Jehn, 1995). Although relationship conflict creates dissatisfaction for the team, it often does not overly disrupt the team's performance. In many cases, team members try to avoid working with members with whom they do not get along personally. Consequently, relationship conflict hurts performance only when the task requires interdependent actions.

This distinction between the effects of task and relationship conflict does not always hold (DeDreu & Weingart, 2003). Conflict disrupts performance and reduces satisfaction because it creates stress and negative feelings and distracts members from performing the task. Low levels of conflict in decision-making tasks may improve the quality and creativity of decision making, but this effect vanishes when the conflict becomes more intense. A little conflict may stimulate thinking, but as conflict intensifies people are distracted by their emotions.

The reality is that task and relationship conflict are often correlated (DeDreu & Weingart, 2003). Disagreement on a task issue can lead to personal attacks. It is difficult to say "I don't like your ideas" and not have it heard as "I think you are stupid." One factor that affects the relationship between task conflict and performance is trust. When team members have a high degree of trust in each other, task-related conflict is less likely to lead to relationship conflict. Teams with high levels of trust can tolerate task-related conflict and use the conflict productively.

Teams can benefit from task conflict when they have a high degree of trust and psychological safety (DeDreu & Weingart, 2003). To use conflict constructively, teams need to cultivate an environment that is open and tolerant of diverse viewpoints, where team members feel free to express their opinions and have the ability to resist pressure to conform to the group (Ilgen et al., 2005). They need to develop cooperative work relationships so that disagreements are not misinterpreted as personal attacks. A "constructive controversy" uses communication styles that focus on issues and ideas and not personal criticism (Tjosvold, 1995).

4. Conflict Resolution Approaches

The conflict resolution approaches available to teams vary, depending on the team members' desire to be assertive and cooperative. Because team members have long-term relationships with one another, they should try to use a collaborative approach to conflicts whenever possible.

Two Dimensions of Conflict

There are several ways people and teams can try to resolve conflicts. The approaches they take depend on their personalities, their social relations, and the particular situation. The types of conflict resolution approaches can be analyzed using the following two dimensions: distribution (concern about one's own outcomes) and integration (concern about the outcomes of others) (Rahim, 1983; Thomas, 1976; Walton & McKersie, 1965). In other words, people in a conflict can be assertive and try to get the most for themselves, or they can be cooperative and concerned with how everyone fares. These two dimensions are independent and lead to the creation of five different approaches to conflict resolution (Figure 7.1).

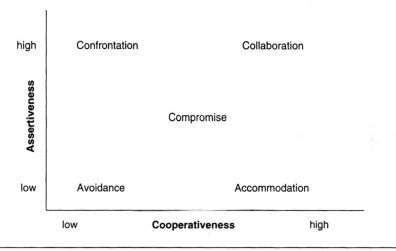

Figure 7.1 Conflict Resolution Approaches

SOURCE: Used by permission of Marvin Dunnette.

1. Avoidance: This approach tries to ignore the issues or deny that there is a problem. By not confronting the conflict, team members hope it will go away by itself.

2. Accommodation: Some team members may decide to give up their position in order to be agreeable. They are being cooperative, but it costs the team the value of their opinions and ideas.

3. Confrontation: Acting aggressively and trying to win is one way to deal with a conflict. However, winning can become more important than making a good decision.

4. Compromise: One way in which to balance the goals of each participant and the relations among the teams is for everyone to "give in" a little.

5. Collaboration: When both sides of a conflict have important concerns, the team needs to search for solutions that satisfy everyone. This requires both cooperativeness and respect for others' positions.

Comparing Different Approaches to Conflict Resolution

Although all these approaches can be used to resolve conflict, each approach has problems. Avoidance, accommodation, and confrontation all may work to resolve the conflict, but these approaches create winners and losers. Teams using these styles often have trouble implementing decisions and find themselves addressing the same issues later. Compromise works somewhat better because everyone wins a little and loses a little. A compromise promotes equity or fairness, but usually does not result in optimal decisions.

When possible, teams should use a collaborative approach to conflict resolution. In collaboration, team members search for the alternative solution that allows everyone to win. Although finding a collaborative solution may be time-consuming and difficult, it has many benefits. Collaboration encourages creativity, leads to greater commitment to decisions, and improves relationships among team members (Pruitt, 1986).

Members of work teams have long-term concerns about their relationships that go beyond specific situations or conflicts. Teams with shared goals and long-term commitments are likely to show more concern for other team members when conflicts arise. Conflicts arise from different perspectives and interests, but the shared goals encourage concern for the perspectives of others. This is why work teams tend to use collaboration and accommodation when resolving internal conflicts (Farmer and Roth, 1998).

Although collaboration may be the best approach in theory, it cannot always be achieved in practice. Occasions arise when different approaches to conflict resolution are best. For example, in a conflict with an emotionally upset boss, a good short-term strategy is to be accommodating. In an emergency situation, people are more likely to accept a confrontational style because they value a quick resolution. Collaboration is the best approach when team members have relatively equal status and there is time to work through a solution. Other approaches may be better when there are large differences in power and a quick resolution is needed.

5. Negotiation, Mediation, and Arbitration

When a conflict becomes intense, the team is sometimes unable to resolve it alone. When teams have conflicts with each other, conflict resolution may be difficult because there may be no personal connection to encourage cooperation. In such cases, outside help may be required to resolve the conflicts. For example, when Boeing was building the 777, it had numerous design teams working on various parts of the jet. To help resolve the inevitable differences among these teams, the company established a set of "integration" teams whose function was to mediate differences among the design teams (Zaccaro & Marks, 1999).

Negotiation

Negotiation or bargaining is the process by which two sides engaged in a dispute exchange offers and counteroffers in an effort to find a mutually acceptable agreement. One of the most important dimensions in understanding how negotiation works in conflict resolution is whether participants have a win-win or a win-lose perspective (Walton & McKersie, 1965). A win-lose perspective is based on the belief that what is good for one side is incompatible with what is good for the other, and that the other side places the same importance on issues that one's own side does (Thompson & Hastie, 1990). With a win-win perspective, participants believe a solution that satisfies both sides is possible. A negotiation can adopt either a win-lose approach and try to reach a compromise or a win-win approach and search for an integrative solution.

In a win-lose approach, there are two types of tactics participants can use successfully. The first approach is to stand tough (Siegel & Fouraker, 1960). The stand-tough approach works well, especially when there is external pressure to make a deal. However, the stand-tough approach may lead to a stalemate or participants may walk away from the situation. In the second approach, called GRIT (graduated and reciprocal initiative in tension reduction), the participants initiate small concessions, reward reciprocal concessions, and build a relationship on the basis of reciprocal trading (Osgood, 1962). This approach is especially useful if the parties in the conflict must work together in the future.

The win-win approach, called integrative bargaining, looks for a creative solution that satisfies the needs of both sides (Pruitt, 1981). People are more satisfied than with the compromise approach and the relationship is better with an integrative solution. However, the integrative approach is more difficult to apply and often requires skillful negotiating.

For example, a budget committee may have limited resources to allocate to two programs. Supporters of each program will forcefully argue for their positions because the overall funds available are fixed. A compromise (splitting the money equally) might leave each program with less money than it needs to operate. An integrative solution may combine the programs' administrative activities to reduce costs, allowing sufficient resources to fully operate both programs.

Mediation and Arbitration

Negotiation is when two parties bargain to try to resolve a conflict. Mediation and arbitration are when a third (outside) party intervenes to help resolve a conflict. In mediation, the intervener does not have authority to impose a solution but acts as a facilitator. Arbitration is a third-party intervention in which the intervener may impose a solution. The intervener acts as a judge.

The advantage of third-party mediation is that it enables conflicting parties to make concessions in a negotiation and save face (Pruitt, 1981). The main tactic is to allow the parties to reorient the situation from win-lose to win-win. The parties can then focus on what they really want from the conflict, rather than thinking about how to beat the other side (Thompson & Hastie, 1990). One of the advantages of mediation is that the mediator has the skills to help both sides work through the issues.

Mediators operate by gaining trust among participants, managing hostilities, developing solutions to conflicts, and gaining commitment to the solutions from participants. They use a variety of tactics to do this. Some tactics focus on the emotional or relationship aspects of situations, whereas others are oriented more toward problem-solving aspects. Mediators may search to develop creative win-win solutions, or they may focus on pressuring the parties into accepting compromise solutions (Carnevale, 1986).

Third-party approaches work, but there can be problems with using them in place of negotiation. Mediation requires voluntary compliance to be effective; the parties cannot be forced to accept a mediator. If either of the parties in a conflict does not want mediation or does not want to settle the conflict, the mediator is unlikely to be successful (Hitrop, 1989).

Sometimes the threat of arbitration will encourage participants in a conflict to reconcile their differences. However, if their differences are substantial, the threat of arbitration may cause each side to harden its position in preparation for arbitration (Pruitt, 1986). People sometimes prefer not to use arbitration because they lose control over the solution or they suspect the arbitrator of being biased. Commitment to arbitrated agreements is often weaker than commitment to negotiated ones (Thomas, 1992).

Whether mediation is preferable to arbitration depends on the situation (Pruitt & Carnevale, 1993). Arbitration is best when there is time pressure and the issues are complex. Mediation is better when there is a relationship among the parties that will continue in the future.

6. Managing Team Conflicts

The goal of managing team conflicts is to develop integrative agreements beneficial to both sides. Integrative agreements are more rewarding than compromises and improve ongoing relationships among parties (Pruitt, 1986). The keys to developing integrative agreements are focusing attention on interests rather than positions and developing trust and rapport between the conflicted parties.

When a team becomes involved in a conflict, members often form coalitions on the basis of positions toward the conflict. Rather than focusing on issues of interest to them, they focus on whether others are for or against their positions. Thompson and Hrebec (1996) found that 50% of people in a conflict failed to realize when they had interests completely compatible with each other, and 20% failed to reach agreement even when their interests were compatible. One reason for this failure is that they did not exchange information about their interests and overlooked areas of common interest (Thompson & Hastie, 1990).

Successful conflict management requires developing trust among participants (Ross & Ward, 1995). If a member trusts members of the other side and believes they want a fair solution, he or she is better able to negotiate a solution. Many conflict reduction approaches are designed to build trust among the parties in a conflict. For example, in a study of bargaining through e-mail, allowing participants to engage in a "get acquainted" telephone call before the bargaining session increased chances of reaching an agreement by 50% (Nadler, Thompson, & Morris, 1999).

Imagine being on a committee whose goal is to reduce violence in local high schools. As the committee begins to search for solutions, a conflict arises over whether the schools should use electronic surveillance technology. Committee members divide into sides over this issue, and all future ideas are evaluated on the basis of support or opposition to this position. Over time, the debate becomes increasingly hostile; new ideas are rejected according to whoever expressed them, rather than being evaluated for their quality.

The solution to this conflict is to find an integrative agreement that addresses the committee's goal (i.e., to improve safety in the schools) but does not depend on either position. The participants need to step back from their emotional involvement in the support of their positions and understand what is really important to them. There are alternative approaches to reducing violence without decreasing privacy in the schools. Some examples include training students in conflict management, using students to monitor compliance with safety rules, and providing teachers with training to help them deal with aggressive incidents.

The search for an integrative solution can be difficult. It is often useful for a team to either use an outside facilitator for a difficult conflict or receive training in facilitating conflicts. The following (adapted from Fisher, Ury, & Patton, 1991) is the structure for negotiation of a conflict:*

1. Separate the people from the problem.
   - Negotiations must deal with both the issues and the relationship, but these two factors should be separated.
   - Diagnose the cause of the conflict. What goals are in conflict? Identify what each side in a conflict wants; make sure each side clearly understands the issues.
   - Encourage both sides to recognize and understand their emotions. Ask them to view the conflict from the perspective of the other side and practice active listening.

2. Focus on the shared interests of all parties.
   - Focus on the issues, not on positions.
   - Identify how each side can get what it wants. Determine the issues that are incompatible between the two sides. Recognize that both sides have legitimate multiple interests.
   - Have each side identify and rank its goals in the conflict. This often shows that the important goals of each side are different, thereby helping each side to see how to trade off unimportant goals to get what it really wants.

3. Develop many options that can be used to solve the problem.
   - Creatively try to generate alternatives that provide mutual gains for both sides. Separate generation of ideas from selection of alternatives.
   - Look for areas of shared interest. Invent multiple solutions as well as solutions to parts of the problem.
   - Practice viewing the problem from alternative perspectives.

4. Evaluate the options using objective criteria.
   - Develop objective criteria to use as a basis for decisions. Define what fair standards and fair procedures to use to resolve the conflict. Agree on these principles before agreeing on a solution.
   - Talk through the issues in order to eliminate unimportant issues. Discuss important differences, searching for the common points on each side.
   - Focus on solutions to which both sides can agree. Do not give in to pressure.

---

*Adapted from Fisher, R., Ury, W., & Patton, B. (1991). *Getting to yes: Negotiating agreement without giving in (2nd ed.).* Boston: Houghton Mifflin.

5. Try again.
   - Creative solutions are difficult to develop. Practice equals success.
   - Teams do not always resolve their conflicts, but they do try to manage conflicts while working through their various tasks.
   - Establish monitoring criteria to ensure that agreements are kept.
   - Discuss ways in which the team can deal with similar issues in the future. How can the team improve its ability to manage conflicts?

Although teams often react spontaneously to conflicts when they occur, teams can engage in preemptive conflict management to help avoid conflicts (Marks, Mathieu & Zaccaro, 2001). Preemptive conflict management strategies include the development of cooperation and trust-building among members, team contracts that identify how to handle difficult situations, and the development of norms for managing communications within a team. These actions reduce the destructive impact of conflicts when they occur.

Summary

Conflict is a normal part of a team's existence. It is a sign of healthy team interactions. However, teams often do not handle conflict well. They make bad decisions in order to avoid conflict rather than learning how to manage it effectively.

Conflict may be analyzed in terms of its sources and types. Conflicts that are healthy for a team come from disagreements on how to address task issues; conflicts that are unhealthy originate from organizational, social, or personal sources. The type of conflict determines the way it should be managed. When conflicts are about misunderstandings and task issues, they can be managed using negotiation to develop acceptable agreements. When conflicts arise from social or personal sources, they often require team building to develop social skills and improve social relations.

Conflict brings both benefits and problems to a team. Conflict helps the team perform its task by fostering debate over issues and stimulating creativity. Conflict hurts the team when it creates strong negative emotions, damages group cohesion, and disrupts the team's ability to operate.

Approaches to resolving conflicts vary depending on how assertive participants are about getting their way and how cooperative they want to be. Although team members use different conflict resolution approaches depending on the situation, collaboration typically is the most effective approach. Collaboration attempts to identify an alternative solution that satisfies both parties. Although they may be more difficult and time-consuming to achieve, collaborative solutions encourage acceptance and support for the solutions.

When a conflict persists, a team may use a more formal approach to conflict resolution. A team typically tries to resolve conflict through negotiation or bargaining. One of the most important factors in bargaining is whether the participants view the situation as a win-lose or a win-win scenario. Mediation and arbitration involve the use of a third party to resolve a conflict. These approaches may be useful for difficult-to-solve conflicts, but the third party must be accepted by all participants in order to be effective.

Improving a team's ability to manage conflict can be achieved in several ways. Developing trust provides a foundation for resolving team conflicts. Learning how to fight constructively is a useful skill. Conflict mediation facilitators have developed helpful approaches for structuring the negotiation process.

TEAM LEADER'S CHALLENGE—7

The high school in your town has been having problems. Recently the number of gangs at school has increased, and acts of vandalism and juvenile delinquency are also increasing. Although there have not been any major outbreaks of violence, stories in the media of violence in other communities have raised concerns among parents. The school board has created a committee of teachers, administrators, students, and concerned parents to develop proposals for dealing with problems at the local high school. You are the leader of this committee.

The meetings started with polite sharing of ideas, but tensions soon became apparent. The four groups had very different ideas about the degree of seriousness of the problem and appropriate solutions. Polite criticism of ideas shifted into cynical asides and finally into heated attacks. As people became more emotional, the negative comments became more personal. You are aware that some of the participants have fought over other school issues in the past.

How do you reduce the negative emotions in this situation?

Should you focus on the social/emotional aspects or the task/policy issues?

What can be done to negotiate agreement among the four groups?

ACTIVITY: OBSERVING CONFLICT RESOLUTION STYLES

Objective. Team members use one of the following five styles to handle conflicts and disagreements. Avoidance means trying to ignore the issue or deny that there is a problem. Accommodation means giving up one's position in order to be agreeable. Confrontation means acting aggressively and trying to get one's way. Compromise means seeking a balance so that everyone gets part of what they want. Collaboration means searching for a solution that satisfies everyone.

Activity. Observe a team or group discussion and note what happens when there is conflict or disagreement. As an alternative, divide a group and assign them positions on a debate topic. Or, the Team Leader Challenge presents a conflict with multiple roles (teachers, administrators, students, and parents) that could be assigned and used to create a conflict. Classify team members' responses to a conflict or disagreement using the conflict resolution styles listed above. Using Activity Worksheet 7.1, note the frequency of each conflict resolution style you observe.

ACTIVITY WORKSHEET 7.1
Observing Conflict Resolution Styles

	Team Members					
	1	2	3	4	5	6
Avoidance						
Accommodation						
Confrontation						
Compromise						
Collaboration						

Analysis. Which conflict resolution style did the team use most often? Did certain team members adopt a similar style for each conflict? How effective were the conflict styles in persuading others? Did the team handle its conflicts in a constructive manner?

Discussion. How can the team better handle conflicts? What can be done to encourage more use of collaboration as a conflict resolution style?

8

Power and Social Influence

G roups use their power to influence behaviors by providing information on how to behave and exerting pressure to encourage compliance. Team members gain power from personal characteristics and their positions, and use a variety of power tactics to influence other members. The dynamics of power in teams is a major influence on leaders' behaviors, how team members interact, the impact of minorities, and the amount of influence members have on one another.

Empowerment is at the core of teamwork where members have been given power and authority over a team's operations. Within the team, members need to learn how to use their own power to work together effectively. Learning how to act assertively, rather than passively or aggressively, encourages open communication and effective problem solving.

Learning Objectives

1. Understand how conformity and obedience influence people's behaviors.

2. What are the different bases of power?

3. How does one decide which influence tactics to use?

4. How does having power change the power holder?

5. How does unequal power affect team interactions?

6. What makes a minority influential?

7. What is empowerment?

8. What problems does an organization encounter when trying to empower teams?

9. How do passive, aggressive, and assertive power styles affect a team and its members?

1. Definitions of Power and Social Influence

Social influence refers to attempts to affect or change other people. Power is the capacity or ability to change the beliefs, attitudes, or behaviors of others. We often think about power in terms of how individuals try to influence one another, but a group has collective power. Conformity occurs through influence from the group, either by providing information about appropriate behavior or through implied or actual group pressure. In addition, obedience occurs through influence from the leader or high-status person in the group.

There is an important distinction between compliance and acceptance. Compliance is a change in behaviors due to pressure or influence, but it is not a change in beliefs or attitudes. Acceptance is a change in both behaviors and attitudes due to social pressure. However, if individuals are repeatedly influenced to change their behaviors, they often internally justify the new way of behaving. Therefore, changing behaviors often leads to changes in attitudes.

Why do people change because of social influence? Social psychologists provide two main reasons for the effects of social influence: normative influence and informational influence (Deutsch & Gerard, 1955). Normative influence is change based on the desire to meet the expectations of others and be accepted by others. Informational influence is change based on accepting information about a situation from others.

Social psychologists have conducted several classic studies on power to demonstrate the basic characteristics of social influence and show some factors that affect the influence process. These studies show how a team influences the behaviors of its members and the power team leaders have over members.

Conformity

Asch's (1955) conformity studies show that even when group pressure is merely implied, people are willing to make bad judgments. The participants in these experiments were asked to select which line was the same length as a target line. Participants who worked alone rarely made mistakes. However, when participants were in a room with people giving the wrong answers, the participants gave the wrong answers 37% of the time. Only 20% of participants remained independent and did not give in to group pressure. The others conformed to group pressure even though there was no obvious pressure to conform (i.e., no rewards or punishments).

Follow-up studies using this approach to study conformity helped explain why people gave in to the group even when there was no direct pressure. For many of the participants, the influence was informational; they reasoned that if the majority were giving answers that were obviously wrong, then the participants must have misunderstood the instructions. Other participants went along with the majority for normative reasons. They feared that group members would disapprove of them if their answers were different. Later studies showed that nonconformists were rated as undesirable group members.

The level of conformity is affected by the group size and unanimity. A group of about five people shows most of the conformity effects. There is not much difference in conformity when using larger groups (Rosenberg, 1961). Unanimity is very important. Many of the conformity effects are greatly reduced with limited social support for acting independently (Allen & Levine, 1969).

These studies show the power a team has over its members. In these experiments, temporary groups set up in psychology laboratories were able to change what people believed and how they behaved. The impact on a team where members have ongoing relationships with one another is much stronger conformity. This is especially true when the team has a high degree of group cohesion; cohesive groups have more power to influence members (Sakuri, 1975).

Obedience

The Milgram (1974) obedience studies show that people are obedient to authority figures even when the requested behaviors are inappropriate. In these obedience studies, participants believed they were part of a learning experiment. They were asked to give an electric shock to a learner whenever the learner made a mistake. They also were told to increase the level of shock with each mistake. Nearly all participants were willing to administer mild shocks; most (65%) continued to administer shocks even after they had been informed that the learner had a heart condition or the learner had stopped responding, and they could see that the shocks being administered had increased to dangerous levels.

The level of obedience in these studies was influenced by several factors. The more legitimate the authority figure, the more likely people were to be obedient. They were more likely to obey when the authority figure was in the room monitoring their performance. Whenever possible, participants did not shock the learner and lied to the authority figure about it. The closer the participants were to the victim (and could see or hear the victim's pain), the less obedience there was. Finally, when there was a group of people running the shock device, participants were less obedient if one other person refused to obey.

The important finding in the Milgram studies is that obedience occurs even when the authority figure does not have power to reward or punish participants. In most teams, the leaders are given limited power by their organizations. For example, team leaders usually do not conduct performance evaluations of members; evaluations usually are done by outside managers. Even without this source of power, the tendency of team members to obey authority figures gives leaders considerable power over team operations.

2. Types of Power

Team members use various types of power to influence one another and the team. The types of power that members possess can be examined in several ways. The study of bases of power is concerned with the sources of power, whereas the study of influence tactics examines how various power tactics are used.

Bases of Power

There are two types of power that an individual can have in a group or organization: personal or soft power, and positional or harsh power (French & Raven, 1959; Raven, Schwarzwald, & Koslowsky, 1998). Personal or soft power derives from an individual's characteristics or personality and includes expert, referent, and information power. Positional or harsh power is based on an individual's formal position in an organization. It includes legitimate, reward, and coercive power. Definitions for these bases of power are provided in Table 8.1.

The types of power are related to each other and often used together (Podsakoff & Schriesheim, 1985). For example, the more one uses coercive power, the less one is liked, so one has less personal or soft power. The more legitimate power one has, the more reward and coercive power one typically has. Because team leaders have less legitimate power than traditional managers, they often rely on expert and referent power to influence the team (Druskat & Wheeler, 2003).

The use of the personal sources of power is often more effective than the use of positional sources (Kipnis, Schmidt, Swaffin-Smith, & Wilkinson, 1984). One reason for this is that the targets of influence are more likely to

TABLE 8.1

Types of Power

Personal / Soft Power

Expert	Power based on one's credibility or perceived expertise in an area.
Referent	Power based on another's liking and admiration.
Information	Power based on the knowledge or information one has about a topic.

Positional / Harsh Power

Legitimate	Power based on the recognition and acceptance of a person's authority.
Reward	The ability to reward (reinforce) a desired behavior.
Coercive	The ability to threaten or punish undesirable behavior.

SOURCE: Adapted from French, J., & Raven, B. (1959). The bases of power. In D. Cartwright (Ed.), *Studies in social power* (pp. 150–167). Ann Arbor: University of Michigan Press.

resist the use of positional power and are less satisfied with its use. Because of this, leaders typically prefer using expert power most often and coercive power least often. However, the use of expert power is limited. The fact that someone is an expert in one area does not make him or her an expert at everything.

Reward and coercive power can be used to influence people to do what is desired, but people do it only because of the reward or fear of punishment. The result is compliance but not acceptance. These strategies are useful for changing overt behaviors, but not for changing attitudes and beliefs; the influencer has to monitor the behaviors to ensure that results are forthcoming (Zander, 1994).

Teamwork should rely on the personal power of team members. Group decision making is better when people who are most expert or have relevant information to add dominate the discussion, rather than when people who have the authority to make decisions dominate. Cooperation is more likely to be encouraged by using personal power sources than by using threats of punishment by team leaders. When team leaders rely on positional power to get their teams to comply with their requests, members are likely to feel manipulated and may resist.

Influence Tactics

Team members can use a variety of social influence tactics to influence one another. Descriptions of these tactics are presented in Table 8.2. Their use depends on the target for influence (e.g., subordinate, peer, superior) and the objective of the influence (e.g., assign task, get support, gain personal benefit) (Yukl & Guinan, 1995).

TABLE 8.2

Social Influence Tactics

Rational argument	Use of logical arguments and factual information to persuade.
Consultation	Seek a person's participation in the decision.
Inspirational appeals	Attempt to arouse enthusiasm by appealing to a person's ideals.
Personal appeals	Appeal to a person's sense of loyalty or friendship.
Ingratiation	Use of flattery or friendly behavior to get a person to think favorably of you.
Exchange	Offer to exchange favors later for compliance now.
Pressure	Use of demands, threats, or persistent reminders.
Legitimizing tactics	Make claims that one has the authority to make the request.
Coalition tactics	Seek the aid and support of others to increase power of request.

SOURCE: Adapted from Yukl, G. (1989). Managerial leadership: A review of theory and research. *Journal of Management, 15,* 251–289.

These power tactics vary by directness, cooperativeness, and rationality. Direct tactics are explicit, overt methods of influence (e.g., personal appeals and pressure), whereas indirect tactics are covert attempts at manipulation (e.g., ingratiation and coalition tactics). Cooperative tactics encourage support through rational argument or consultation; competitive tactics attempt to deal with resistance through pressure or ingratiation (Kipnis & Schmidt, 1982). Finally, some tactics are based on rational argument or the exchange of support, whereas inspirational and personal appeals rely on emotion.

People prefer direct and cooperative strategies. The most effective tactics are rational argument, consultation, and inspirational appeals (Falbe & Yukl, 1992). These are the more socially acceptable tactics and are useful in most situations. However, status differences in a group determine which tactics are used. Traditional leaders often use pressure and legitimizing tactics on subordinates, while subordinates often use rational argument, personal appeals, and ingratiation to influence leaders. Team leaders may try to reduce status differences in the team by using more cooperative influence strategies (Druskat & Wheeler, 2003).

3. Power Dynamics

The use of power changes the dynamics of the group process. Unequal power changes the way the leader treats other team members and the way members communicate with one another. Subgroups that disagree with the majority can have substantial influence on how the team operates. The level of interdependence among team members changes the power they have over one another.

Status and the Corrupting Effect of Power

Power is rewarding, so people with power often want more of it (Kipnis, 1976). It has a corrupting influence; people with more power often give themselves a higher share of rewards. It is easy for someone with power to give commands rather than make requests. Because powerful people get mostly positive feedback from subordinates, they begin to care less about what subordinates say and have an inflated view of their own worth.

Kipnis (1976) demonstrates the corrupting nature of power in studies on groups in business organizations and families. He documented a cycle of power where power leads to a desire to increase one's power. Table 8.3 shows how the cycle operates.

TABLE 8.3

Cycle of Power

- Access to power increases the probability it will be used.
- The more power is used, the more the power holders believe they are in control.
- As the power holders take credit, they view the target as less worthy.
- As the target's worth is decreased, social distance increases.
- Use of power elevates the self-esteem of the powerful.

SOURCE: Adapted from Kipnis, D. (1976). *The powerholders*. Chicago: University of Chicago Press.

One of the problems with this effect is that its impact is often unconscious. Over time, powerful leaders come to believe their subordinates are externally controlled, and therefore must be monitored and commanded by their leaders to get them to do anything. It is a self-reinforcing cycle. A team may try to deal with this problem by rotating team leaders. When leaders know they will eventually become just another team member, they are less likely to use controlling power tactics.

Unequal Power in a Team

Groups vary in the ways power is distributed. When groups have unequal power levels among members, there tends to be more mistrust, less communication, and more social problems than in more egalitarian groups. Groups with powerful leaders tend to have less communication and more autocratic decision making, thereby reducing the quality of team decisions.

Unequal power is often caused by status differences, which have an impact on team communication (Hurwitz, Zander, & Hymovitch, 1953). High-status members talk more and are more likely to address the entire group. Members communicate more with high-status people and pay more attention to what they say. Low-status members often talk less and are unwilling to state their true opinions if they differ from those of high-status people. Consequently, when high-status people speak, people either agree or say nothing, so high-status people have more influence in group discussions. This communication pattern does not lead to good decision making or to satisfied and motivated team members.

In theory, a team should have only equal-status communication, but this is not always the case. The team leader may assume a higher status than the other team members. A team is sometimes composed of members with different levels of status within the organization. Team members should leave their external status positions at the door so that everyone on the team has equal status. However, it is difficult to interact as an equal with someone in one situation and be deferential in other situations.

When power is unequal because of status or other factors, a team can try to improve the situation by using group norms to equalize power and control communication. Norms level the playing field in a group. They equalize power by putting constraints on the behaviors of powerful members. For example, the norm of majority rule in decision making limits the power of the leader. The group may have norms that encourage open and shared communication, prevent the use of intimidation or threats, and value independent thinking. All these help reduce the impact of power differences in the group.

Minority Influence

Most of this discussion of power has focused on the impact of powerful people or pressure from the group majority. However, a group may contain individuals or subgroups that resist group pressure. A minority may resist the leader and group pressure, and eventually be influential in changing the group (Moscovici, 1985). The ability of minorities to influence the majority group depends on their consistency, self-confidence, belief in their autonomy, and relationship to the group overall.

Minorities become influential by sticking to their positions (Nemeth, 1979). When minorities are consistent, they make the majority think about its position. A group can put quite a bit of pressure on minorities to change, so it takes self-confidence to resist this pressure. Minorities must appear to be autonomous and able to make their own choices if they are to be influential. If the minorities are viewed as being supported or influenced by an outside group, their impact is reduced. Finally, minorities must appear to be part of the group. They are less effective if they reject the group or are always seen as dissenters (Levine, 1989).

It can be difficult to be a minority team member who disagrees with the majority. Because of the desire to be accepted by the team, individuals are often unwilling to disagree or even present an alternative view of an idea. When minority opinions have some support within the team, they are more likely to be expressed and accepted by the team (Ilgen et al., 2005). Teams that create a climate of trust and psychological safety encourage members to express their unique opinions.

An important value of a minority is its ability to stimulate team members to view an issue from multiple perspectives (Peterson & Nemeth, 1996). When a minority disagrees with the team's view of a situation, the team is encouraged to rethink its position and generate more alternative views. The overall effect of this is to encourage more flexible thinking, which increases creativity and innovation. The minority may not get its way, but over time it will have a substantial impact on how the team thinks and acts.

Impact of Interdependence

Task interdependence is the degree to which completing a task requires the interaction of team members. Teams with high levels of interdependence are more likely to be effective if they have autonomy, or the power and authority to control how they operate (Langfred, 2000). In highly interdependent teams, autonomy allows team members to work together more efficiently, control their own interactions, and increase internal coordination. These actions help improve performance. In teams with low levels of interdependence, the team members are accustomed to working independently, so increases in communication and coordination do not help improve performance.

Interdependence may help a team perform better by changing the amount of power team members have over one another (Franz, 1998). Dependence in a relationship is one of the bases of power. Heightened levels of overall task interdependence are associated with increased personal power. The more team members need one another to complete a task, the more power each team member has over the group.

4. Empowerment

Empowerment in a workplace refers to the process of giving employees more power and control over their work. It is the shifting of power and authority from managers to employees. In one sense, empowerment is the core notion of teamwork. A team must have the power to control how it operates; this is what makes a team different from a work group. A team cannot operate successfully if a manager controls its internal operation, or if its external relationships are completely controlled by an organization.

The success of empowerment programs depends on the organization's willingness to share information and power with its employees (Hollander & Offerman, 1990). Leaders love the idea of empowerment in theory, but they primarily engage in command and control actions because that is what they are comfortable doing (Argyris, 1998). Delegating authority is stressful for team leaders because organizations typically hold the leader responsible for the outcome of the teamwork. Although they may feel discomfort delegating authority to a team, supporting empowerment leads to more effective team performance.

Team empowerment has four dimensions: potency, meaningfulness, autonomy, and impact (Kirkman & Rosen, 1999). Potency is the collective belief that the team can be effective. Meaningfulness refers to the team experiencing its work as valuable and worthwhile. Autonomy is the degree to which the team controls the way it works and experiences independence. Impact refers to whether the team's work is important for the organization.

Empowerment is a benefit both to the individuals and teams that are empowered and to their organizations. Employees who work in empowered jobs have increased motivation and job satisfaction (Ford & Fottler, 1995). Empowerment increases employees' confidence in their ability to perform a task. The ability to change how the team works encourages continuous improvements and innovation (Burpitt & Bigoness, 1997). Empowered teams provide better customer service because they are more willing to accept responsibility for handling customer problems (Kirkman & Rosen, 1999). Organizations benefit by having teams that function more effectively, have greater organizational commitment, and show increased acceptance of change.

Degrees of Empowerment Programs

The amount of empowerment a team possesses depends on the actions of external leaders, the responsibilities given to the team, the organization's human resources practices, and the social structure of the team (Kirkman & Rosen, 1999). External leaders encourage team empowerment by allowing the team to set its own performance goals and letting the team decide how to accomplish those goals. When teams are given more responsibilities for production, customer service, or quality improvements, they experience more empowerment. Human resources policies may give the team more control over staffing, performance evaluation, and training to promote empowerment. Increased participation in team decision making promotes empowerment.

Organizations use a variety of approaches to promote empowerment, ranging from simple changes (e.g., suggestion boxes for employee input) to work teams, to fully empowered self-managing teams (Lawler, 1986). Although these programs have features in common, they differ in what is shared and the breadth of involvement activities.

Sharing information is the minimum requirement for empowerment. However, for a team to feel fully empowered, the power to make and implement decisions is necessary. Without some power to act, employees have little incentive to continue to improve the way their team operates.

A second dimension of analyzing empowerment programs is the breadth of empowerment activities. Most empowerment programs give team members control over job content (i.e., the task and work procedure they perform) but not over job context (i.e., goals, reward systems, and personnel issues) (Ford & Fottler, 1995). For example, quality programs may allow employees to make changes to improve the quality of their work operations but may not allow employees to influence personnel decisions. By contrast, at the Saturn automobile plant, which uses empowered work teams, team members control their work processes, deal with external customers, hire new team members, and conduct performance evaluations.

Successful Empowerment Programs

Although research shows that empowerment can be effective, few organizations are willing to make the transition because of their beliefs about power. If power is viewed as a limited commodity, giving power to a team reduces managerial power. This ambivalence about power makes managers reluctant to empower teams (Herrenkohl, Judson, & Heffner, 1999). However, when teamwork programs are successful, everyone in the organization gains in power.

One of the main problems with empowerment programs is resistance from managers and supervisors (Lawler & Mohrman, 1985). Managers are often told to empower work teams under their supervision, but they are held responsible for how the team performs. This is why surveys show that although 72% of supervisors believe that empowerment is good for the organization, only 31% believe that it is good for supervisors (Klein, 1984). Although some supervisors support empowerment, many are concerned about loss of status and lack of support from upper management.

Resistance from supervisors and managers may be handled in a number of ways (Klein, 1984). Supervisors need to be involved in the design of the empowerment programs. The roles of supervisors should be evaluated, and their new responsibilities and authority should be clearly defined. Often supervisors need additional training in teamwork skills to prepare them for their new roles as team leaders. Additional training has a secondary benefit in that it demonstrates the organization's commitment to its supervisors.

5. Application: Acting Assertively

People express power through their behaviors. They may act passively, aggressively, or assertively (Alberti & Emmons, 1978). Their emotional tones and the ways they confront problems define these power styles. Assertiveness is both a skill and an attitude (Jentsch & Smith-Jentsch, 2001). When team members act assertively, they show their willingness to be independent and accept responsibility for their actions.

The use of these power styles has important impacts on communication in teams. For example, teamwork problems are one of the chief causes of major airline accidents. Accidents happen when crewmembers are unwilling to communicate problems to their superiors. This lack of assertiveness in a team is a major problem for airline crews, medical teams, and police and firefighting teams, which is why assertiveness training is a standard element in training for action teams (Cannon-Bowers & Salas, 1998).

When people talk about power styles, there is some confusion whether the styles are an element of someone's personality, a behavior, or something in between. Power styles are like personality traits; some people adopt preferred styles and use them in most situations. However, most people change styles depending on the situation, and people can be taught how to use particular styles. This suggests that power styles are more like behaviors than personality traits. Table 8.4 presents an overview of these power styles, their impacts on teams, and situations in which they often are used.

TABLE 8.4
Power Styles

	Styles	Impact	Use
Nonassertive/ Passive	Polite and deferential; avoid problems	Resentment and confusion	Dangerous situations; unequal status
Aggressive	Forceful and critical; focused on winning	Satisfaction and withdrawal	Emergencies; unequal status
Assertive	Clear and confident problem solving	Satisfaction and trust	Most situations equal status

Nonassertive/Passive. The nonassertive or passive style is polite and deferential. A sweet, pleasant, or ingratiating emotional tone is added to one's communication. A person using the passive approach tries to avoid problems by not taking a stand or by being unclear about his position. By being evasive, the person using this approach is trying not to upset or anger anyone by disagreeing. The person's desire to be liked by others is based on his personal insecurity or fear of the situation.

The goal of the passive approach is to win approval and be liked. Unfortunately, this style does not work well. Passive people often feel stressed and resentful, in part because problems never seem to go away. The receivers of passive communication often have mixed responses. On the one hand, they get their way; on the other, they are uncertain what the passive communicator really believes, and lack respect for him or her.

The passive approach is appropriate in some situations. When a conflict becomes highly emotional, a passive response may defuse the situation. When interacting with a person of higher status, a passive response may be expected from a subordinate. Acting aggressively toward the boss may be an inappropriate or dangerous tactic. There are situations in which being assertive is too risky.

Aggressive. The aggressive style is forceful, critical, and negative. A negative emotional tone is added to the communication so that it appears more powerful. A person using the aggressive approach deals with problems and conflicts by trying to win and refusing to compromise. The underlying emotions of the aggressive style are anger, insecurity, and lack of trust. In some ways, this is similar to the passive style, which is why people sometimes swing back and forth between passiveness and aggressiveness without ever being assertive.

Use of the aggressive style is often rewarded. In many situations, people give in to people who are acting forcefully, in part because of a misapplication of the rule of reciprocity. People may let someone have his or her way if it is important to him or her, because they expect to get their way in things that are important to them. What the aggressive person does is act forcefully in all situations. There is a cost to the aggressive style. People on the receiving end of the aggressive style feel resentment, act defensively, and try to withdraw from the situation.

The aggressive style can be appropriate, as in problem or emergency situations requiring forceful action. It can be a valuable approach for dealing with blocked situations where progress is stalled. If there is resistance and change is vital to the group, aggressiveness may be the appropriate response.

Assertive. The assertive style uses clear and confident communication. No emotions are added to messages. Assertiveness is communicating openly with concern for both others and oneself. It is taking responsibility for one's own communication. The assertive person takes a direct problem-solving approach to conflicts and problems. The goal is to find the best solution, so the person is willing to listen and compromise.

Although they are not always successful, assertive communicators are generally satisfied with and relaxed in their performances. The assertive style shows respect and encourages trust in others and open communication in a team. High self-esteem and trust in the group underlie the assertive style.

The assertive approach is appropriate in most situations in which people are interacting on an equal basis; therefore, it should be the most typical type of communication within a team. The absence of the assertive approach is a sign of unequal status differences that are disrupting communication, or unresolved conflicts that are creating a defensive communication environment.

Use of Power Styles

Teams are more productive when their communication is primarily assertive (Lumsden & Lumsden, 1997). Both passive and aggressive styles create resentment and inhibit open communication. Teams may adopt unproductive power styles for several reasons. For example, power styles trigger other power styles. An aggressive style triggers a passive response, whereas an assertive style triggers an assertive response. Assertiveness is a power style used among people with relatively equal power or status; passiveness and aggressiveness are power styles used among people with differing statuses or levels of power.

Differences in power styles are often attributed to personality, gender, or racial differences. This explanation is rarely true and gets in the way of improving team communication. For example, women often act more passively in business teams, causing some men to assume that women are more passive. Kipnis (1976) shows that this stems from organizational power, not gender. When men and women have equal power in a situation, women are no more likely to act more passively.

Assertiveness is primarily a reflection of the distribution of power in a group. To encourage assertive communication, the group needs to reduce power differences among members. In an organization, people work in a hierarchy that gives everyone different amounts of power. When people are working in a team, however, they need to treat fellow team members as equals.

Encouraging Assertiveness

Assertiveness is the power style most appropriate for teamwork. The primary key to encouraging assertiveness is to equalize power among team members. However, equalizing power might not be enough. People develop habits in communicating, so it may be necessary to provide them with training in assertive communication (Alberti & Emmons, 1978). Assertiveness training programs use a number of techniques to encourage better communication in teams:

1. Active listening: Active listening is summarizing and repeating a speaker's message to ensure that it is understood. This technique clarifies the message, shows respect and attention, and encourages more communication.

2. Positive recognition: Learning how to give others positive recognition reduces the need for manipulative power tactics. Too often, high-status people criticize what they do not like but never acknowledge what they like. Positive recognition acknowledges someone's work and helps encourage more of what is desired in the performance.

3. Clear expectations: Learning to state expectations clearly is another communication technique that encourages assertiveness. People often misinterpret behaviors as inappropriate or resistant when they are really caused by not understanding what is desired. Clarifying expectations lets everyone know what the issues are.

4. Saying no: For people who overuse the passive style, learning how to say no to inappropriate requests is an important skill. Passive communicators often feel guilty when they turn down requests. This encourages them to agree to do something and then passively resist doing it.

5. Assertive withdrawal: Being assertive is not always the right response, so people need to know when not to participate. When situations become too emotionally heated or threatening, people need to learn how to send a clear message of their desire to postpone or terminate conversations.

Assertiveness is a situation specific behavior (Jentsch & Smith-Jentsch, 2001). The willingness to be assertive depends on the situation and the individual's relationship to the other people involved. People may be assertive in social or personal situations, but not in work situations. They may be assertive with friends, but not with strangers or business associates. This is why it is important to conduct assertiveness training in environments where assertive behavior will need to be applied.

Summary

Power is the ability to change the attitudes, beliefs, and behaviors of others. Groups have power because they can influence members by suggesting how people should behave. Groups also exert social pressure to get members to conform to group norms. Group leaders have the power to influence others through obedience to authority.

Power can be analyzed by examining where it comes from and the types of techniques that are used. Group members gain power through personal bases (e.g., being an expert) and positional bases (e.g., having authority delegated by the organization). Influence tactics can be based on encouraging others or trying to control others. People prefer to use personal power bases and cooperative tactics because these approaches are less likely to create resistance.

The use of power by groups has several important dynamics. Power tends to corrupt its users. People with power tend to use it and take personal credit for the success of their groups, thereby encouraging them to use power more often. Unequal power in teams caused by status differences among members disrupts group communication. High-status people talk more, people tend to agree with high-status people, and low-status people become reluctant to state their true opinions. Minorities in groups may be influential if they are consistent, self-confident, and autonomous. By resisting the influence of the majority, minorities focus their group's discussion on their positions. Interdependence among team members increases their power to influence one another.

One of the core notions of teamwork is empowerment. To be effective, teams need power, authority, and responsibility to control their own behaviors. Unfortunately, organizations often have trouble sharing managerial power with teams. Empowerment programs may range from simple information-sharing with team members to development of self-managing teams. Successful empowerment programs must deal with the perceived loss of power by supervisors and managers by incorporating them into the teams' activities.

Team members may act in a passive, aggressive, or assertive manner. These are personal power styles. Although there are situations in which acting passively or aggressively is appropriate, in most group situations assertiveness is the best approach. Assertiveness encourages clear communication and a rational approach to problems, but is disrupted by unequal status in the group. There are several techniques for training group members to act more assertively.

TEAM LEADER'S CHALLENGE—8

You are the manager of a technical services team that provides support services for other organizations. Over the past year, you have been trying to make the transition to a more self-managing team. Now you call yourself the "team leader" rather than the manager. As part of the transition, the team now makes decisions about scheduling, partnering, and other task assignments. Team meetings where employees passively listened to instructions have been replaced by team discussions and group decision making. You have worked hard to encourage team members to speak up at meetings, and they are contributing more now.

At today's meeting, a team member suggested a new way to organize work practices. You told the team you didn't like the idea because you tried a similar plan in the past and it did not work well. Another team member ignored your explanation and complained that you were stifling innovation and were unwilling to share power with the team.

Should the team leader allow the team to make a mistake in order to learn?

What is the best way to handle team members who challenge the leader's authority?

How much power sharing (or empowerment) is appropriate for a work team?

ACTIVITY: USING POWER STYLES—PASSIVE, AGGRESSIVE, AND ASSERTIVE

Objective. Team members may use three different power styles. The nonassertive or passive style is polite and deferential. This approach tries to avoid problems by not taking a stand or by being unclear about one's position. The aggressive style is forceful, critical, and negative. This approach deals with problems and conflicts by trying to win. The assertive style uses clear and confident communication. The assertive person takes a direct problem-solving approach to conflicts and problems.

Activity. Observe the interactions in a team or a group discussion of the Team Leader Challenge. Write down examples of passive, aggressive, and assertive behaviors you have observed. Using Activity Worksheet 8.1, note the frequency of team members acting passively, aggressively, or assertively. An alternative activity is to break into groups and have individuals take turns using one of the three power styles during a group discussion. Participants should analyze their own and others' performances according to their perception and reactions to each style.

Passive behavior: _____

Aggressive behavior: _____

Assertive behavior: _____

ACTIVITY WORKSHEET 8.1

Observing Passive, Aggressive, and Assertive Power Styles

	Team Members					
	1	2	3	4	5	6
Passive						
Aggressive						
Assertive						

Analysis. Which power style did the team use most often? Did certain team members adopt a similar power style for most communication? Was the team's communication dominated by assertive or by passive or aggressive communication? Did the leader primarily use the assertive style?

Discussion. What triggers the use of a power style? Is it personality or team characteristics? How would you encourage the team to engage more in equal-status assertive communication?

9

Decision Making

Decision making is a central activity of teams. One of the greatest benefits of teams is their ability to bring together multiple skills and perspectives in making decisions. Teams use different approaches to make decisions, from consultation to consensus. These approaches vary in speed, quality, and acceptance by team members.

Teams encounter a number of problems when trying to make good group decisions. Group polarization and "groupthink" are two examples of these problems. There are structured decision-making approaches that help improve the decision-making process. Although it may be difficult at first, learning how to make consensus decisions is an important skill for teams to develop.

Learning Objectives

1. What are the main advantages and disadvantages in using groups to make decisions?
2. What factors make group decisions superior to individual decisions?
3. How are consultative, democratic, and consensus decision making different?
4. What factors are useful for evaluating a decision-making approach?
5. How does the normative decision-making theory help teams make decisions?
6. What factors disrupt the ability of groups to make good decisions?
7. How do group polarization and groupthink affect teams' decision-making process?
8. What are the benefits of and problems with structured decision-making techniques such as the nominal group technique?
9. How can teams improve their ability to achieve consensus decision making?

147

1. Value of Group Decision Making

Using groups to make decisions creates both advantages and disadvantages for teams. Depending on the situation, group decisions may or may not be superior to individual decisions.

Advantages and Disadvantages
With Group Decision Making

A group brings more resources to a problem than are available to one person. Group members pool their knowledge through group discussion; their interaction leads to new ideas that no single member would have developed (called process gain). Also, incorrect solutions are more likely to be identified and rejected. A group has a better memory for past facts and events, so it is less likely to repeat mistakes. Overall, group members combine different skills and knowledge to make higher-quality decisions.

Group decision making has motivational effects on group members (Zander, 1994). Being part of a group encourages members to try to make good decisions and perform better. They are more committed to a decision in which they participated, so are more likely to support its implementation.

Group decision making affects the skills of group members and the team as a whole. Members benefit by gaining a better understanding of the issues involved by participating in the discussion. These benefits are lost if decisions are forced on the group by the leader or an outside source. The team also benefits by learning to make decisions. Over time, a team can become more efficient at decision making, thereby reducing many of the problems with group decision making.

The main disadvantage with group decision making is that groups are less efficient in making quick decisions because they suffer from process loss (Steiner, 1972). When groups enter into discussions, some of the discussions are about coordination and social issues. This "wasted" discussion time prevents groups from focusing solely on their tasks.

Groups encounter many communication problems when trying to make decisions (DiSalvo, Nikkel, & Monroe, 1989). To be efficient, group decision making requires skillful facilitation, and some leaders and groups lack these skills. Decisions can get bogged down in emotional conflicts that waste time and damage the morale of groups. Powerful team members or people who like to talk too much can dominate discussions and disrupt groups' ability to make decisions. Finally, discussions can get sidetracked or interrupted, or become disorganized.

One of the benefits of group discussions is the availability of information from the variety of experiences and skills of group members. However, group discussions are not necessarily good at incorporating this information. Analyses of group interactions show that information known by most of the group members is more likely to be discussed than is information held by specific individuals (Stasser & Titus, 1985). In other words, groups do not pool all the knowledge available; they focus on the knowledge that is common to all members.

Finally, sometimes a group can work hard to make a decision when it is not really important (Zander, 1994). The group may be asked to make a decision, but it is really only a recommendation, and the decision is to be made higher in the organization. This creates a sense of wasted time and effort and may discourage future participation in the group.

When Are Group Decisions Superior to Individual Decisions?

Group decisions work better than individual decisions when groups successfully pool resources to solve problems or make decisions. Successful pooling is affected by several factors. One is group composition. Groups with heterogeneous members with complementary skills make for superior group decisions. Diversity of opinion is a major advantage of using groups (Wanous & Youtz, 1986). If a group is composed of similar members with identical skills and knowledge, there is little benefit from making group decisions.

Another factor is good communication. Group decisions work better only if the discussion process successfully pools the knowledge and ideas of the group members. However, group discussions typically focus on shared information rather than the unique information held by members. Poor communication skills and problems managing group discussions can prevent groups from using their resources (DiSalvo et al., 1989).

A third factor relates to the need for groups to make decisions. Groups are needed for tasks that are too complex for one individual to perform or problems that are too difficult for one individual to solve. For a simple problem, the issue is whether anyone has the correct answer and whether the group will accept the correct answer. A simple problem does not require that the group spend time making a decision.

From these considerations, one can outline the types of situations in which individual decision making is better than group decision making. Individual decision making is preferable when the issue does not require action from most group members, the decision is so simple that coordination is not needed to implement it, or the decision has to be made quickly.

2. Approaches to Group Decision Making

When thinking about how to make decisions, groups often decide to vote. However, this can be a problem. There is a variety of methods groups can use to make decisions, and it is important to use the type of decision-making process best suited to the problem. Not all problems need full participation, and voting can sometimes create problems rather than solve them. For important decisions, groups might need to reach consensus.

The options that teams can use to make decisions may be viewed as lying along a continuum, from leader-based decisions to decisions made with full participation of the team (Johnson & Johnson, 1997). These options are shown in Table 9.1. Although there are many approaches a team can utilize, teams typically use either consultative, democratic, or consensus decision making.

In consultative decision making, one person has authority to make the decision, but he or she may ask for advice and comments from team members before deciding (Kerr & Tindale, 2004). Although this advice does have an

TABLE 9.1

Approaches to Group Decision Making

Leader Oriented	Leader decides
	Leader assigns expert to make the decision
	Consultative—leader consults with team and then decides
Group Technique	Mathematical techniques (averaging)
	Structured decision techniques (e.g., nominal group technique)
	Democratic—voting with majority rules
Full Participation	Consensus

impact, the leader typically gives more weight to his or her own opinion and the opinions of members with similar views.

The consultative approach is often used in a work group when the leader has management authority and responsibility for the group decision. Teams also use consultative decision making when a project is divided into parts and one person is responsible for a part. That person may ask for advice and may need to coordinate with others, but if it primarily affects the person's part of the project, he or she makes the final decision.

The consultative approach uses only some of the team's resources. Its disadvantages are that it does not fully develop commitment to a decision, does not resolve conflicts among team members, and may encourage competition among team members to influence the leader. However, it is a very efficient decision-making approach, taking little time to complete, and it may provide the leader with crucial information to help make a decision.

In democratic decision making, the groups votes on a decision. One of the major advantages of the democratic approach is that it is a quick way of including all team members' opinions. Majority decisions often work better than stricter criteria (such as two-thirds or unanimity) (Kerr & Tindale, 2004). Simple majorities produce high-quality decisions with little cognitive effort.

Although voting is a popular decision-making style, it can create problems for a team. Voting can prematurely close discussion on an issue that has not been fully resolved. This can lead to a lack of commitment from the losing minority. Because there are winners and losers, voting may create resentment among team members. Those who disagree with the vote may be unwilling to support and implement the decision after it has been made (Castore & Murnighan, 1978). It can be useful for decisions that have limited conflict and implementation issues.

The consensus approach to decision making requires discussion of an issue until all members have agreed to accept it. Acceptance does not mean that the decision is a member's favorite alternative. It means that the member is willing to accept and support the decision.

Consensus decision making might be time-consuming, but it is the best way to fully use team resources. When it is successful, it also improves the team operation. The consensus approach should be used for important decisions requiring the full support of the team for implementation. It takes time, energy, and skill to reach consensus, but consensus decisions have a greater likelihood of being implemented by the team.

In a sample of work teams in more than 100 companies, Devine, Clayton, Philips, Dunford, and Melner (1999) found that most (62%) of the teams used consensus decision making. In the other teams, decisions were made by the leaders or managers (25%) or by voting (13%). The use of consensus decision making was positively related to team effectiveness for all types of teams.

Evaluating Group Decision-Making Approaches

The primary criteria for evaluating a decision-making approach are quality, speed, and acceptance or support (Johnson & Johnson, 1997). A good decision-making approach should use the resources of the team to make a high-quality decision. The decision-making process should reflect efficient time management. Once a decision has been made, the members of the team should be willing to accept it and support its implementation. The importance of these three criteria varies depending on the problem or situation. For example, in an emergency situation, time is essential and acceptance is less important because people are often willing to support any decision.

In general, decision-making techniques that include group discussion and participation lead to higher-quality decisions; this is especially true if the problems are complex or unstructured, or if leaders do not have enough information to make good decisions. However, for some issues, good leaders can make high-quality decisions alone. The importance of quality as a decision criterion also varies. Some issues are relatively trivial, so high-quality decisions are not necessary. In such cases, having leaders make the decisions saves time for groups.

Group decision making is slower than individual decision making, but the importance of speed as a criterion varies. In many cases, the issue is not speed, but rather prioritizing the decisions a team needs to make. Some decisions are important and must be made quickly, whereas other decisions should be put off until the team gathers more information. Teams often spend too much time on unimportant decisions and not enough time on the important ones.

The third factor is acceptance. To the extent that a decision requires the support and acceptance of team members to be implemented, the decision should include input from team members (Murnighan, 1981). Teams often use decision-making techniques (e.g., voting, averaging) to speed up the decision-making process, but these techniques can limit the level of acceptance of the decisions. When acceptance is important, teams should use consensus decision making.

These three evaluation criteria are interrelated. The relationship between quality and speed is fairly obvious. When more time is available for making a decision, there is also more time for gathering and analyzing information, which improves the quality of the decision. Speed and acceptance are also related. Comparisons of Japanese and U.S. decision making found that U.S. organizations make decisions faster. However, the U.S. organizations are both slower and more likely to fail at implementation. In Japanese organizations, final decisions are not made until commitment has been gained to implement the decisions from all relevant participants. Once they make decisions, they are quickly able to implement them.

Normative Decision-Making Theory

How should a team leader choose to make a decision? There are both advantages and disadvantages with the consultative, democratic, and consensus approaches. It is difficult to sort out these factors and determine the best approach for a team. Normative decision-making theory addresses this problem (Vroom & Jago, 1988; Vroom & Yetton, 1973). It is a leadership theory that can be used by teams to help select the best decision-making approach.

Normative decision-making theory is based on the assumption that the best type of decision-making approach depends on the nature of the problem. The problem determines how important time, quality, and acceptance are in reaching a decision. Once the nature of the problem is understood, the best decision-making approach can be selected. This practice is typically used by the leader to analyze a problem to determine the best decision-making approach.

The analysis of the problem focuses on two issues: whether a quality decision is important, and whether acceptance of the decision by subordinates is important. Seven questions are used to analyze a problem (see Table 9.2). Once a leader analyzes the nature of the problem, a decision tree is used to tell the leader what type of decision-making approach to use. In general, the leader should use more group-oriented approaches (e.g., democratic, consensus) when a high-quality decision is needed or when team acceptance

---

TABLE 9.2

Questions for Analyzing a Problem

1. Is a high-quality decision required?

2. Do I have enough information to make such a decision?

3. Is the problem structured?

4. Is it crucial for implementation that subordinates accept the decision?

5. If I make the decision alone, is it likely to be accepted by my subordinates?

6. Do subordinates share the goals that will be reached through the solution of this problem?

7. Do subordinates disagree about appropriate method for attaining goals so that conflict will result from the decision?

SOURCE: Adapted from Vroom, V., & Yetton, P. (1973). *Leadership and decision making*. Pittsburgh, PA: University of Pittsburgh Press.

is needed to implement the decision. Research testing of the theory has been fairly supportive, although the results are dependent on the decision-making skills of the leader. (It does not work to tell a team leader to use consensus decision making if the leader does not have the skills to facilitate the decision-making process.)

The normative decision-making theory makes some important points about group decision making. When the decision is important and requires support to implement, the decision-making process should be group oriented. However, when the decision is trivial and just needs to be made, it is a waste of the team's time to discuss it. Often people believe that democratic or consensus decision making is best for value reasons, but this can lead to a team wasting too much time on trivial issues. One important function of the team leader is to manage the situation. The leader handles the minor and administrative decisions so that the team has the time to focus on the important issues.

3. Decision-Making Problems

There are many different types of problems that can disrupt a team's ability to make a good decision. Disagreements, time pressure, and external stress can cause these problems. Group polarization can affect a group decision by making the result more extreme due to interpersonal processes. Groupthink describes a number of group decision-making flaws caused by the group's desire to maintain good relations rather than to make the best decision.

Causes of Group Decision-Making Problems

Disagreements. Probably the most common group decision-making problem is premature closure; that is, trying to avoid disagreement by voting to make a quick decision. This technique works for making the decision, but often leads to implementation problems later. Politics, a domineering leader, hidden agendas, poor norms, and other factors can cause disagreements. Many of these problems relate to the group process rather than to the topic of the decision. When these problems disrupt the discussion of the decision, the group needs to focus on improving its internal communication.

Too little disagreement can also be a problem. Disagreement helps stimulate thinking and leads to better decisions. Group discussions with some disagreement lead to better decisions than do conflict-free group discussions (Schwenk, 1990). However, these constructive conflicts come at a cost. Group discussions with substantial disagreement are rated as less satisfying experiences by group members and reduce interest in continuing to interact with the group.

Time Pressure to Decide. Groups respond to time pressure by trying to make quick decisions. To do this, they often use decision-making approaches that are simple and inadequate (Zander, 1994). For example, a group may support the first useful suggestion and ignore or prevent further discussion of alternatives. A group may select a plan that has worked in the past without fully examining whether it is applicable to the current situation. Finally, a group may delegate the decision to the leader or a group member, thereby forgoing the benefits of group analysis and decision.

Outside Stress. Stress from forces outside a team may lead to poor decision making. These outside forces could be external competition, upper management, or other parts of the organization. When groups experience stress, they have a stronger desire for uniformity of opinion among members (Kerr & Tindale, 2004). This desire means that the team will exert stronger pressure on deviant opinions. This can lead to groupthink or allowing the leader to make the decision. Stress disrupts the decision-making process by reducing the number of ideas generated and the analysis of issues. It can cause a desire to make decisions more quickly in order to reduce uncertainty, thereby rushing the decision-making process. All this leads to poor-quality decisions.

Group Polarization

Although it might be expected that the outcome of group discussion would be a decision that is close to the average of the group's initial position, this is not always the case. The effect of a group discussion is to make the final decision more extreme than the average of the members. This may be either more risky or more cautious, depending on the initial inclination of the group. This phenomenon is called group polarization.

Original research by Stoner (1961) showed that groups made riskier decisions than did individuals. This was called the risky shift phenomenon. However, subsequent research showed that this actually was an intensification effect. Groups tend to move toward an extreme and become either more risk-oriented or more conservative (Myers & Lamm, 1976). The group polarization effect occurs only when the group has initial tendency agreement, not when there are major differences of opinion among the members. There are several explanations for group polarization that examine the role of normative and informational influences.

Normative influence describes how the existing group norm affects the decision-making process. Group members want to create a favorable impression, so they compare their answers to the group's norm and then shift their positions to be more consistent with it (Myers & Lamm, 1976). The group norm shifts as members change their positions in an attempt to be more typical of the group's position. The combined effect of these shifts is to move the group decision more to the extreme. This is especially important when the decision is primarily a matter of values or preferences.

Information influence is caused by the amount of exposure to information during a group discussion. When a group discusses an issue, most of the discussion is from the dominant position (Kerr & Tindale, 2004). People prefer to hear information they agree with, so other group members reward the sharing of common information. Because group members are more exposed to arguments supporting the dominant position, they shift their opinions in that direction (Burnstein & Vinokur, 1977).

Groupthink

One of the most famous types of group decision-making problems is groupthink, a term coined by Janis (1972). Janis used the analysis of historical decisions, such as the Cuban missile crisis of the 1960s, to show how decision-making processes can go wrong. Groupthink occurs when group members' desire to maintain good relations becomes more important than reaching a good decision. Instead of searching for a good answer, they search for an outcome that will preserve group harmony. This leads to a bad decision that is accompanied by other actions designed to insulate the group from corrective feedback. Since the initial identification of groupthink, researchers have expanded on the causes and implications of the phenomenon (Table 9.3).

Three main factors contribute to groupthink: structural decision-making flaws, group cohesiveness, and external pressure (Parks & Sanna, 1999). Structural decision-making flaws create bad decisions because they impair the group decision-making process. These flaws include ignoring input from outside sources, a lack of diversity in viewpoints within the group, acceptance of decisions without critical analysis, and a history of accepting decisions made by the leader. Group cohesiveness encourages groupthink by creating an environment that limits internal dissension and criticism. Pressure from outside for a decision limits discussion time and encourages the group to support the first plausible option presented to the members.

The external pressure experienced by the group leads to a set of symptoms of groupthink. These symptoms convince the group that it has made a good decision and that everyone in the group agrees with it. Consequently, there is internal pressure on members not to voice their concerns and objections. The collective effect of these symptoms is a poor decision, made without considering alternative options or longer-term consequences of the decision.

TABLE 9.3

Model of Groupthink

Antecedent Conditions	Structural—domineering leader and limited input from outside the group. Cohesiveness—desire to maintain good relations is dominant. Stress—outside forces put stress on the group to make a decision.
Symptoms	Illusion of invulnerability—belief that the group's decision will work. Collective rationalization—group members supporting one another's ideas. Belief in morality of the group—belief that group's actions are inherently right. Direct pressure on dissenters—suppression of negative comments in group discussion. Stereotypes of out groups—overestimation of the group's superiority compared with others. Self-censorship—group members do not state their opinion if it differs from the group's. Illusion of unanimity—belief that everyone agrees with the decision. Self-appointed mind guards—group members protect the leader and group from negative information about the decision.
Decision Defects	Group considers only a few alternatives when making a decision. Group fails to examine the adverse consequences of its decision. Alternatives are eliminated without careful consideration. Group does not seek the advice of outside experts. Group does not consider what to do if the decision does not work.

SOURCE: Adapted from Janis, I. (1972). *Victims of groupthink.* Boston: Houghton Mifflin.

A number of techniques have been proposed to prevent groupthink. A group may create the role of critical evaluator to comment on the group's decisions and processes. Group members may be asked to discuss alternatives with experts outside the group, especially experts who might not agree with the direction in which the group is going. After a decision has been made, the group should schedule another meeting to examine remaining doubts about it.

4. Decision-Making Techniques

Several decision-making techniques have been developed to manage group decision-making problems. These techniques structure the decision-making process with a set of process rules. They are technically good approaches, but they sometimes appear like magic to the users. An answer appears that represents the group's opinion even though the group has never discussed the issues.

Nominal Group Technique

The nominal group technique is a decision-making technique that allows a group of people to focus on the task of making a decision without developing any social relations. It is called nominal because it does not require a true group; it can be used by a collection of people who are brought together to make a decision.

When using this technique, the leader states the problem to the group. People write down their solutions to the problem privately. Each person then publicly states his or her answer, and the answers are recorded so that everyone can see them. Group members may ask questions to clarify the others' positions, but they cannot criticize the ideas. The participants then use a rank-ordering procedure to rate the value of the solutions. This rank ordering is used to select the group's preferred solution.

The advantage of the nominal group technique is that it is relatively quick, discourages pressure to conform, and does not require group members to get to know one another before the decision-making process (Delbeq, Van de Ven, & Gustafson, 1975). However, it requires a trained facilitator to conduct it, and it can address only one narrowly defined problem at a time.

Delphi Technique

The Delphi technique uses a series of written surveys to make a decision (Dalkey, 1969). A group of experts is given a survey containing several open-ended questions about the problem to be solved. The results of this survey are summarized and organized into a set of proposed solutions. This is returned to the participants, who are then asked to comment on the solutions from the first survey. The process is repeated until the participants start to reach agreement on a solution to the problem.

The Delphi approach is useful when it is necessary to include a specific set of people in a decision, but those people are distributed geographically and cannot meet in a group (Delbeq et al., 1975). The number of people involved makes no difference, so a large group of people could participate at the same time. This approach is also useful when there is great disagreement on an issue that requires subjective judgments to resolve. However, the process is time-consuming (more than a month for a typical decision) and requires skills in developing and analyzing surveys.

Ringi Technique

The Ringi technique is a Japanese decision-making technique used for dealing with controversial topics (Rohlen, 1975). It allows a group to deal with conflict while avoiding a face-to-face confrontation. (Face-to-face confrontations are considered inappropriate in Japanese culture.) In this approach, a written document presenting the issue and its resolution is developed anonymously. This document is circulated among group members, who individually write comments, edit the document, and forward it to other group members. After completing a cycle, the comments are used to rewrite the document, and it is recirculated through the group. This process continues until group members stop writing comments on the draft.

The Ringi approach can be a slow process, and there is no guarantee that the group will come to agreement. However, anonymous comments allow everyone to state their true convictions, while avoiding embarrassing others in a confrontation.

Evaluation of Decision-Making Techniques

These techniques structure the group decision-making process to eliminate all but task-oriented communication among group members. The Delphi and Ringi techniques use only written communication. The nominal group technique requires each group member to generate ideas independently and then to interact only to choose among alternative ideas. This structuring of the decision-making process allows decisions to be made by larger groups of people who do not have to meet. These approaches can produce decisions that are as good as, or better than, decisions produced from group discussions, and they can do so with fewer productivity losses. In addition, with these approaches, people are satisfied with their levels of participation (Van de Ven & Delbecq, 1974).

However, these techniques are based on the assumption that socializing, unequal participation, and other aspects of group discussions are problems. Not all group researchers agree on this point (McGrath, 1984). The social aspects of the group process may have significant benefits. The detached and impersonal atmosphere of these decision-making techniques reduces people's acceptance and commitment to a decision. In addition, the political acceptability of the solution may not be as sensitive to the organization as a solution produced by a group discussion, which typically is dominated by the higher status participants.

5. Application: Consensus Decision Making

Consensus decision making uses all of a team's resources fully, encourages support for implementation of decisions, and helps build team skills. Consensus decision making is a slow process because people typically are not good at it. A team should practice consensus decision making to improve its decision-making skills so that when important problems arise, the team will have the ability to handle the problems effectively. This is a developmental process; learning to do consensus decision making increases the team's ability to use it.

Reaching consensus does not mean every team member believes the solution is best (Hackett & Martin, 1993). Consensus is achieved when each team member can say yes to the following questions:

- Will you agree that this is what the team should do next?
- Can you go along with this position?
- Can you support this alternative?

In other words, a team can support the decision 100% even though not all members completely agree with it. Consensus is the voluntary giving of consent.

A team leader or facilitator can help gain consensus through a number of techniques. The team needs to be given adequate time to work through an issue. Conflict should be viewed as inevitable, so team members need to be encouraged not to give in just to avoid conflict. Flipping coins or voting when differences emerge is not an acceptable alternative. Team members need to recognize that giving in on a point is not losing, and that gaining is not winning. The goal is to negotiate a collaborative solution, not beat the other side in a debate. Table 9.4 presents some guidelines to help team members reach consensus.

If a group gets stuck trying to reach consensus, it can use several options to break the impasse. The team can agree to not agree, and then move on to a related issue. Changing topics and returning to an issue later reduces the

---

TABLE 9.4

Guidelines to Help Reach Consensus

---

1. Avoid arguing for your own position without listening to the position of others.
2. Do not change your position just to avoid conflict.
3. Do not try to reach a quick agreement by using conflict-reduction approaches, such as voting or tossing a coin.
4. Encourage others to explain their position to better understand any differences.
5. Do not assume that someone must win and someone must lose in a disagreement.
6. Discuss the underlying assumptions, listen carefully to one another, and encourage the participation of all members.
7. Look for creative and collaborative solutions that allow both sides to win, rather than compromises where each side only gets some of what it wants.

SOURCE: From David W. Johnson & Frank P. Johnson. *Joining Together: Group Theory and Group Skills, 6,* reprinted with permission, Pearson Education. Copyright © 2003.

---

emotional tension created during a conflict. If a decision must be made quickly, the team can decide that it must use an alternative such as voting. Or it may decide to develop a compromise solution wherein each side gives in on part of its demands. When time is available, the team may ask for outside help or bring in a trained facilitator to manage the decision process.

Summary

The largest advantage of group decision making is the ability to bring more resources to solving a problem. It also helps motivate team members and develop their skills. However, group decision making takes time and does not always succeed. Group decisions are better than individual decisions when the team has a diversity of perspectives, the discussion is open, and the problem is suitable for a group.

The main approaches to group decision making are the consultative, democratic, and consensus. These approaches vary in time required, quality of the decisions, and support for implementation. The decision-making approach used depends on the nature of the problem. The normative decision-making theory provides a way to analyze problems to help determine the best decision-making approach.

A team's ability to make a decision can be disrupted by too much or too little conflict, pressure to decide quickly, and outside stress. Group decisions tend to be more extreme than individual decisions because of the group polarization effect. The desire to maintain good relations within a group may disrupt the group decision-making process and cause groupthink. Groupthink leads to inadequate decisions that are strongly defended by the group members.

Several structured techniques for decision making have been developed to manage certain problem situations. The nominal group technique may be used in large groups with limited social interaction. The Delphi technique uses a series of surveys for decision making and can be used with large groups that never meet. The Ringi technique is a Japanese approach that helps avoid confrontations in the decision-making process. These techniques gain efficiency by structuring the decision-making process, but they may reduce support for the decision due to a decreased sense of participation.

Consensus decision making is the approach that best uses the resources of a team. A number of techniques can be learned to help improve the team's ability to make consensus decisions.

TEAM LEADER'S CHALLENGE—9

You are the head of a university department with 10 faculty members. Although you try to organize meetings using an agenda, things don't always work out as planned. Faculty members often spend too much time talking about minor issues and do not get around to dealing with important issues. Most of the time, you try to get the group to reach consensus on issues. This is because voting often fails to resolve the conflict over the issue, so you have to discuss and re-decide the issue in the future.

The dean is pressuring you to make a decision about an admissions policy for new students. You have brought the topic up several times in department meetings, but the faculty cannot reach an agreement about what the policy should entail. The dean wants an answer soon or he will impose a policy on the department.

How can you improve decision making at department meetings?

What type of decision style should you use for the admissions policy?

If the department is unable (or unwilling) to make a decision, how should you handle the situation?

ACTIVITY: CONSENSUS DECISION MAKING

Objective. Consensus decision making requires discussing an issue until all agree to accept it. Acceptance does not mean that the decision is the member's favorite alternative; it means the member is willing to accept and support the decision. Learning consensus decision making is an important skill for a team.

Activity. Form a group and have it develop consensus answers to the following questions:

What is the most important skill for a team member to possess?

What is the most important characteristic of a good team leader?

What is the greatest benefit of using teamwork?

What is the greatest problem with using teamwork?

While the group is trying to reach consensus, have an observer use Activity Worksheet 9.1 to note whether the group follows the "Guidelines to Help Reach Consensus" presented in Table 9.4.

ACTIVITY WORKSHEET 9.1
Observing the Guidelines to Help Reach Consensus

Did the team follow the guidelines presented below?	Yes	No
1. Avoid arguing for your own position without listening to the position of others.		
2. Do not change your position just to avoid conflict.		
3. Do not try to reach a quick agreement by using conflict reduction approaches, like voting or tossing a coin.		
4. Try to get others to explain their position so that you better understand the differences.		
5. Do not assume that someone must win and someone must lose when there is a disagreement.		
6. Discuss the underlying assumptions, listen carefully to one another, and encourage the participation of all members.		
7. Look for creative and collaborative solutions that allow both sides to win, rather than compromises.		

Analysis. Was the group successful in reaching consensus for its decisions? Did the group follow the guidelines for consensus decision making?

Discussion. What advice could you give a team to improve its ability to make consensus decisions?

10

Leadership

A team has many ways of selecting a leader and assigning leadership roles. The leader may be assigned by the organization, the team may be self-managing, or leadership roles may be distributed among team members. What is the best style of leadership? There is no definitive answer to this question, but a number of approaches have been suggested. Situational leadership theory is one approach to helping the leader decide the best way to act, depending on the characteristics of the team members.

Organizations are experimenting with new forms of team leadership. In self-managing teams, many leadership functions are turned over to the teams. Self-managing teams provide a variety of benefits, but they require the development of group process skills and social relations to operate effectively. In addition, team leadership requires new skills and responsibilities compared with those of traditional leadership approaches.

Learning Objectives

1. How do leaders vary by the roles they perform and the amount of power they possess?

2. What factors influence who becomes a team's leader?

3. What are the different types of leadership that a team can have?

4. What are the main approaches to studying leadership, and what are their implications?

5. What are some of the substitutes that reduce the importance of leadership?

6. Understand situational leadership theory. How does a leader's behavior relate to the readiness level of the group?

7. What are the benefits of and problems with self-managing teams?

8. What are the functional roles and responsibilities of team leaders?

1. Alternative Designs of Leadership for Teams

Leadership is a process in which an individual influences the progress of group members toward attainment of a goal. Types of leaders vary by method of selection and the roles they are expected to perform. The person who emerges as the leader of a team may not be the one best suited for the role. Although most teams have designated leaders, the use of self-managing teams provides an alternative to the traditional approach.

Characteristics of Team Leadership

Although we normally think of a single individual holding the position of leader, this is not always the case. Rather than talking about leadership as pertaining to any one person, we need to recognize that leadership is a process or set of functions that may be performed by many of a team's members (Day, Gronn, & Salas, 2004). Teams vary in their types of leaders, the distribution of leadership roles, and the power their leaders have.

There are leaderless groups, teams with leaders assigned by their organizations, teams that select their own leaders, and self-managing teams. Most teams have one person assigned to the role of leader. The leader may be selected by the organization and assigned to the team, the leader may be elected by the team, or the position may be rotated among the team members. A team may start out sharing and rotating the leadership functions until a leader emerges from the team's interactions.

Rather than centralizing the roles of leadership, a team's leadership roles may be divided, on the basis of different tasks performed by the team. Many teams use the STAR (situation or task, action, result) structure for distributing leadership functions (Wellins, Byham, & Wilson, 1991). The team's task is divided into specific functions, and the responsibilities for each function are assigned to a team member to perform. Team members may be rotated through the different roles so as to develop the skills within the team. For example, a factory team may divide its task into quality, safety, maintenance, supplies, and administration. Rather than the team leader being responsible for all of these functions, one team member may be responsible for each function.

Leaders vary in the power or authority they possess. When a leader is assigned by an organization, the leader may have the authority to make the team's decisions. It is then up to the leader to decide how team decisions should be made. When the leader is elected or rotated, he or she typically has limited power and serves primarily as facilitator of the group process. A designated leader with organizational power is useful when the task is very complex and structure is needed, when there is significant conflict among team members, or when someone is needed to manage the relationship between the team and other parts of the organization (Lumsden & Lumsden, 1997).

A team leader is not the same as a manager. A manager is given power and authority by the organization over subordinates, whereas a team leader typically does not have this type of power. A manager is responsible for the actions of his or her subordinates, whereas it is the team (and not the leader) that is responsible for the actions of its members. A manager has the authority to make decisions, whereas a team leader facilitates decision making. Finally, a manager is responsible for handling personnel issues (e.g., employee hiring, evaluation, reward), whereas an organization typically does not give a team leader the authority to perform these personnel functions.

Leader Emergence

When no leader is assigned to a group, a leader emerges from the group to coordinate its actions (Hemphill, 1961). The person who becomes the leader may not be an effective leader. Leaders tend to be taller and older than their followers, but these characteristics are unrelated to effectiveness (Stogdill, 1974). Men are five times more likely than women to be group leaders (Walker, Ilardi, McMahon, & Fennell, 1996). However, gender is unrelated to leader effectiveness (Eagly, Karau, & Makhijani, 1995). Studies of military teams show that leaders have more physical ability and better task performance skills (Rice, Instone, & Adams, 1984). Although task skills relate somewhat to leader effectiveness, physical ability has little relationship to effectiveness.

The most important predictor of group selection of a leader is the participation rate (sometimes called the "babble effect"). Group members are more likely to select the most frequent communicator as the leader (Mullen, Salas, & Driskell, 1989). Unfortunately, the quantity of communication is more important than quality for leadership selection. It appears that people who communicate frequently demonstrate active involvement and interest in the group, and this implies a willingness to work with the group members.

Leader prototype theory provides another way to explain the emergence of leaders with characteristics unrelated to effectiveness (Lord, 1985). This theory examines the relationship between the leader and the perceptions of group members. Members have certain implicit notions about what a good leader is like. To the extent that the leader meets these expectations, the leader is more influential. Although the specific traits of good leadership vary in the minds of followers, effective leaders are usually assumed to be intelligent and dedicated and to have good communication skills.

Leadership prototype theory explains some of the problems with the way team members select leaders. Members rely on their prototypes to decide who should be their leaders, but these prototypes of good leaders are not necessarily accurate. For example, gender differences in leadership may be due to the way gender stereotypes relate to prototypes about leaders. The typical female stereotype emphasizes expressive qualities such as emotion and warmth, whereas the typical male stereotype emphasizes instrumental qualities such as productivity and power (Williams & Best, 1990). Although both expressive and instrumental qualities are needed in leaders, members tend to emphasize the importance of instrumental qualities (Nye & Forsyth, 1991). This causes team members to view males as more likely candidates for leadership and to see male behaviors as more important in leaders.

Leadership Options for Teams

The most common form of team leadership is designation of a leader by the organization. This designation gives the leader formal authority and power. However, the leader is not always the most capable person in all areas, so some of the leadership roles may be taken over by other members. For example, a leader may be selected because of his or her technical skills, so other team members may be needed to handle the group process facilitation aspects of team leadership. In a traditional work group, the leader maintains control over most of the group's decisions. Because the work group does not experience autonomy, it often fails to improve productivity, quality, and morale (Stewart & Manz, 1995).

The main alternative to designated leadership is the self-managing team. In this type of team, there is no leader with organizational authority and decisions are made by consensus. The leader is a facilitator who manages the team's decision-making process (and sometimes other administrative tasks) rather than serving as the manager of the group. Self-managing teams became popular during the 1980s as a way to manage factory work teams, and self-management has spread to other types of teams.

Self-managing teams can have two types of leadership: power-building and empowered (Stewart & Manz, 1995). In power-building leadership, the leader is an active, democratic-oriented person who teaches the group team skills and guides team-building efforts. The leader provides structure, helps coordinate the team, and encourages and rewards good performance. Through delegation and democratic decision making, the skills and abilities of the team grow. However, the leader retains control over team behaviors and long-term strategic direction. Under this type of leadership, self-managing teams develop skills and typically improve performance, quality, and morale.

Empowered leadership is a less involved form of leadership. The leader operates as a facilitator but does not control the team's work processes or major decisions. The team is truly self-governing. Even in this situation, there are important leadership roles. The leader models appropriate behavior, promotes learning and skills development, and deals with issues outside the team. This type of leadership is the most effective for improving performance, quality, and morale. However, it requires a highly skilled and well-developed team. Organizations that want to create self-managing teams typically go through the stage of power-building leadership before making the transition to empowered leadership.

2. Approaches to Leadership

Leadership is a topic most people believe to be very important, and it has been the subject of an immense amount of research. However, we do not understand leadership very well. Problems in studying leadership cause this dilemma. The fact that people believe leadership is important does not make it true (Meindl & Ehrlich, 1987). Rather than one, optimal leader, different types of leaders are useful in different situations, and leaders are more important in some situations than in others.

The four historical approaches to research on leadership have different implications for organizations and teams (Table 10.1). The *trait or personality approach* is based on the belief that good leaders have certain characteristics. If this is true, then psychological tests could be used to identify and select good leaders. An alternative is the *behavioral approach*, which defines leadership by the ways leaders act. This approach attempts to determine what good leaders do in order to train people to be good leaders. The *situational approach* questions the necessity of leadership. It attempts to determine when leaders are needed and what factors can substitute for leadership. The final approach is the *contingency approach*, which attempts to combine

TABLE 10.1

Models of Leadership

Model	Implication
Trait or Personality	Use tests to select good leaders
Behavioral	Train people to be good leaders
Situational	Understand substitutes for leadership
Contingency	Link traits or behaviors to situations

personality or behavioral characteristics of leaders with situational characteristics. For example, it may be impossible to say what a good leader does,
but possible to define good leadership in an emergency situation.

Trait or Personality Approach

The trait approach is the oldest model of leadership, with hundreds of
studies conducted during the 1930s and 1940s (Yukl, 1989). It assumes that
good leaders have a certain set of characteristics. If these characteristics are
identified and measured, it should be possible to know how to select good
leaders.

Many traits have been examined, but research has failed to confirm
a strong relationship between traits and leadership (Kirkpatrick & Locke,
1991). More recent research suggests that sets of traits are associated with
good leadership. For example, effective leaders have more drive, honesty,
leadership motivation, self-confidence, intelligence, knowledge of business,
creativity, and flexibility. No one trait can predict good leadership, but effective leaders do differ from typical followers in exhibiting higher levels of
these characteristics overall. The basic problem is that people who are successful leaders in one situation (e.g., business) are not necessarily successful
in others (e.g., politics, religion).

A good example of the problem with the trait approach is the value of
intelligence. Good leaders should be intelligent. This seems like an obvious
statement, but is it true? Are the most intelligent people the best leaders? Are
the smartest U.S. presidents the most effective? Would most college professors make great business leaders? It is true that good leaders tend to be more

intelligent than average, but leaders are not necessarily the most intelligent people in their organizations. In addition, the importance of intelligence varies. In a dictatorship (e.g., the military), intelligence is an important characteristic of good leaders. In a democracy (e.g., local politics), good leaders must be able to easily relate to others and have good communication skills. These communication skills are more important than intelligence.

Motivation is another example of the problem with the trait approach. Successful leaders are motivated, but what type of motivation is important? The difference between successful entrepreneurs of small businesses and managers of large companies is not a difference in the level of motivation; rather, it is a difference in the type of motivation. Successful managers in large organizations have a strong need for power and a moderately strong need for achievement (McClelland & Boyatzis, 1982). The power motivation of such managers is focused on building their organizations and empowering their subordinates, rather than on gaining personal power and control. By contrast, successful entrepreneurs have a high need for achievement and independence and a less strong need for power.

Flexibility, or the ability to adapt to a situation, is considered an important characteristic of good leaders. Obviously, not all situations require the same approach, given that we also like consistency and leaders who stand for things. We malign politicians who are too influenced by public opinion polls, but that is a flexible approach to leadership. Clearly, we do not want too much flexibility in our leaders.

Behavioral Approach

The behavioral approach defines leadership as a set of appropriate behaviors. The goal of this approach is to define how good leaders act in order to train people to be good leaders. Rather than focusing on issues such as intelligence and creativity, most of the research on leader behavior focuses on two issues: decision-making style and task versus social focus.

The decision-making approach has primarily examined the benefits of authoritarian leadership in comparison with democratic leadership. As was noted in Chapter 9, there is no one best way to make a decision. The best type of decision-making approach depends on the situation or problem (Vroom & Jago, 1988).

Research in this area has demonstrated some of the problems and benefits of different decision-making approaches. Democratic leaders tend to create followers with higher morale, job satisfaction, and commitment. However, democratic decision making can be slow, and leaders may be viewed as weak. Autocratic leaders tend to be more efficient decision-makers, but this style can create dissatisfaction and implementation problems among followers.

Behavioral research also examines whether leaders should focus on the tasks or the social relations among the group members (Likert, 1961). Is a leader's primary role to organize and manage the task, or should the leader ensure that social relations are good, group members feel satisfied and motivated, and the group can maintain itself? Research in this area has been contradictory and inconclusive, except for the finding that group members like leaders who show social consideration (Yukl, 1989).

As with the decision-making style issue, the correct answer here depends on the situation. If a team is performing a routine task, the leader should focus on social relations, because the team does not need help with the task. If a project team is working on a difficult problem, a good leader will help the team better understand and work on the task. If a team is capable of self-management, the leader should ignore both task and social issues and focus on concerns outside the team.

One new example of the behavioral approach to leadership is the leader-member exchange model. This approach looks inside a work group to see how leader and subordinates interact (Graen & Uhl-Bien, 1995). Leaders form different types of relationships with subordinates; this creates in-groups and out-groups. In-group members get more attention from leaders and more resources for performing their jobs. Consequently, they are more productive and more satisfied than out-group members. The distinction is made by leaders early in the relationship and is based on little information about the subordinates. Sometimes it is influenced by irrelevant factors such as similarity, personality, and attraction, rather than by actual performance. The importance of this perspective is the recognition that a team contains a variety of individuals. A leader does not treat everyone alike, and a leader may be performing effectively with some team members but ineffectively with others.

Situational Approach

Are leaders really important to the success of teams? When are leaders important? These questions are the basis for the situational approach to leadership. The value of this approach is in understanding the situational factors that affect leadership and its alternatives.

When historians study great leaders, they note the relationship between leaders and situations. Charismatic leaders require situations in which people have important needs and are searching for others to help resolve them (Bass, 1985). The same is true for other historically important leaders; they led during dramatic times.

People often overrate the importance of leaders (Meindl & Ehrlich, 1987). Although leaders may have a strong impact on the success of organizations, in most day-to-day operations their impact is much less. However, leaders are cognitively important for followers. It is difficult to explain the success or failure of organizations, so leadership becomes a simplified explanation for what has happened and why.

One of the chief values of the situational approach is in examining alternatives to leadership or factors that can substitute for leadership. These factors relate to the characteristics of employees, jobs, and organizations (Yukl, 1994). Competent, well-trained, and responsible employees need leaders to a lesser degree. Routine jobs that are highly structured do not require leader supervision. Organizing into teams and developing a cohesive team spirit reduce the need for leaders.

Contingency Approach

The contingency approach is the researcher's answer to the problems with leadership research. If one cannot define the traits or behaviors of good leaders separate from the situation, then leadership theories should combine these factors. However, a good research theory may be difficult to use in practice. Contingency theories are complex and more difficult to understand and apply than other theories.

Contingency theories start by focusing on some characteristic of a situation. Various theories examine the type of task, level of structure, and favorableness of the situation for the leader. The theories then examine some aspect of the leader's personality or behavior, such as interpersonal skills or task orientation. These two sets of factors are linked to show either how the leader should behave depending on the situation or what type of leader would be best given the situation. For example, because of preference or training, some people tend to be autocratic leaders. Autocratic leaders work well in situations where they have considerable power and followers are motivated to comply. This is why the military trains leaders to handle emergency situations forcefully and selects leaders who can act in this way.

Yukl's (1989) multiple linkage model is a contingency theory that relates to leading teams. The theory states that successful performance of a team depends on the following six intervening variables: member effort, member ability, organization of the task, teamwork and cooperativeness, availability of resources, and external coordination. Situational factors both directly influence these variables and determine which variables are most important. The role of the leader is to manage and improve these intervening variables. In the short run, most leader actions are intended to correct problems in these six variables. In the long run, the leader tries to make the situation more favorable by implementing improvement programs, developing new goals and directions, improving relations with the organization, and improving the team's climate.

3. Situational Leadership Theory

From a teamwork perspective, one of the most important leadership theories is situational leadership theory (Hersey & Blanchard, 1993). This theory links the leader's behavior to characteristics of the group. The value of this theory goes beyond simply telling the leader how to behave. Situational leadership theory is a developmental theory that assumes that one of the goals of leadership is to develop the group. As such, it is the most team-oriented of the leadership theories.

Situational leadership theory starts with the assumption that there are four basic styles of leadership, based on a combination of task and people orientation. Leaders can be directing (high task-oriented), coaching (high task- and people-oriented), supporting (high people-oriented), or delegating (neither task- nor people-oriented). The appropriate style depends on the readiness level of the group. Group readiness is based on the skills of group members, their experience with the task, their capacity to set goals, and their ability to assume responsibility. The connection between the leader's behavior and the group's readiness level is shown in Figure 10.1.

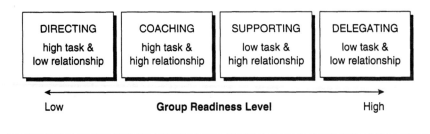

Figure 10.1   Situational Leadership Theory

SOURCE: Adapted from Hersey and Blanchard (1993).

To see how this theory works, imagine being the leader of a group of adolescents in a summer work program. On your first day as leader, your group has little experience with the task and little experience working together. As the leader, you need to take control of the situation and get the group to start working together. A task-oriented (directing) approach is needed. As the group gains some experience, your leadership style should soften (coaching) so as to reward the groups' accomplishments. Once the group learns how to perform the job and act responsibly, you need to further reward members by allowing them to participate in the decision-making process (supporting). This both helps increase their commitment to the group and develops their leadership skills. When the group can take full responsibility for performing its task, your job shifts to addressing issues outside the team, since you are no longer needed to guide the team. As the leader, you delegate most of the internal leadership functions and let the group manage itself.

As may be seen from this example, situational leadership theory makes two important points. First, the leader needs to adjust his style of acting relative to the readiness of the group. Second, leadership is a developmental process, and the leader's behavior should promote group readiness.

4. Self-Managing Teams

As organizations become more team-oriented, they sometimes shift to the use of self-managing teams (Wellins & George, 1991). Self-managing teams provide a number of benefits beyond the use of standard work teams. However, developing self-managing teams can be a difficult process, and this type of team is not suited for all situations.

Self-managing teams shift responsibility for performance to team members (Hackman, 1986). This reduces the need for managers and allows the remaining managers to focus on tasks outside the team. When they are successful, self-managing teams encourage the empowerment of employees and development of team member skills.

Although the idea of self-managing teams has been around since the 1960s (as part of sociotechnical systems theory), the use of self-managing teams was not common until the 1980s, when they were used primarily with factory and service teams (Manz, 1992). By the 1990s, more than 40% of large companies in the United States were using self-managing teams with at least some employees (Cohen, Ledford, & Spreitzer, 1996). The most important reasons for companies to introduce self-managing teams in manufacturing were to improve performance and quality (de Leede & Stoker, 1999).

The main impact of self-managing teams is to shift responsibilities from management to team members. This is not an all-or-nothing process. Rather, there are many levels of self-management, depending on how willing the organization is to give the team responsibility and authority. Table 10.2 shows four different levels of responsibility that a factory team could possess.

TABLE 10.2

Team Empowerment Levels

Level of Responsibility	Team Tasks
Level 1: 20% of job responsibility	Maintenance Quality control Continuous improvement of process
Level 2: 40% of job responsibility	Managing supplies Customer contact Hiring new team members
Level 3: 60% of job responsibility	Choosing the team leader Equipment purchase Facility/work area design
Level 4: 80% of job responsibility	Budgeting Personnel (performance appraisal, compensation, etc.) Product modification and development

SOURCE: Adapted from Wellins, R., Byham, W., & Wilson, J. (1991). *Empowered teams*. San Francisco: Jossey-Bass.

As the team is given more power or responsibility, it takes on more tasks that were formerly handled by managers. When the team reaches the fourth level, it is performing 80% of the tasks formerly performed by managers.

One problem with self-managing teams of factory and service workers is resistance from middle managers and professionals who fear that increased use of teams will reduce the need for their jobs (de Leede & Stoker, 1999). This makes it difficult for self-managing teams to develop good working relationships with other parts of their organizations. Often, technical support functions, such as engineering, are more accustomed to dealing with supervisors than with workers. In addition, some leadership functions are difficult to replace with teamwork, especially those dealing with external relations and personnel issues.

Leading Self-Managing Teams

Leading self-managing teams requires new approaches to leadership (Druskat & Wheeler, 2003). External team leaders often manage several teams and work at the interface between these teams and the organization. These leaders focus on both the team and the organization in their boundary-spanning actions. Their primary functions are to build relationships, search for needed information, gather resources and support from the organization, and empower their team.

Even though self-managing teams control their internal operations, external leaders do have important roles; they provide valuable coaching, support, and motivation for teams (Morgeson, 2005). However, many self-managing service and production teams perform relatively routine tasks and under normal conditions do not require assistance from external leaders. It is when the teams encounter novel events or problems that external leaders may be needed.

Traditional leaders often use commands and take over when there are problems. For self-managing teams, the external leader may either provide support or directly intervene in the team's operations (Morgeson, 2005). When leaders intervene by preparing teams for change or providing supportive coaching, they are viewed as effective leaders and increase satisfaction with leadership; when they directly intervene into the team's operations, they decrease satisfaction with leadership. Active involvement by the leader is viewed as negative by the team, since it takes away the team's autonomy. Active interventions are only related to team effectiveness and satisfaction when the team encounters disruptive events they cannot manage on their own.

Success of Self-Managing Teams

The success of self-managing teams depends primarily on the organizational context (Cohen et al., 1996). Context plays a key role in the level of support teams secure from their organizations, the teams' access to relevant information, and the teams' ability to act independently. The more power and autonomy organizations give to self-managing teams, the greater impact the teams have on improving their performance and the success of their organizations (Cohen & Bailey, 1997).

The benefits of self-management for production and service teams do not necessarily apply to professional and managerial teams. Cohen and Bailey's (1997) review of the factors that relate to team success in different types of work teams found that self-management did not improve the performance of project teams. The highest-performing project teams had leaders who were highly involved in managing the task. Because members of project teams already have substantial autonomy, they may not view self-management as a personal benefit. In addition, their projects are non-routine and difficult, so leaders who help provide structure are viewed as a benefit rather than as unnecessary supervisory control.

There are a number of reasons why production or service teams are better suited for self-management than professional or managerial teams.

In production teams, team members can be cross-trained, which allows them to understand the issues involved in each other's work. In professional teams, team members have different types of expertise or represent different parts of their organizations, thereby limiting members' understanding of each other's perspectives (Uhl-Bien & Graen, 1992). Production teams have clear performance measurements and can use quantitative feedback to evaluate and improve performance. In most cases, it is difficult to analyze and measure the performance of professional teams (Orsburn, Moran, Musselwhite, Zenger, & Perrin, 1990). Finally, production workers place more importance on their social relations and are less competitive when compared to professionals and managers (Lea & Brostrom, 1988).

In their study of professional teams, Levi and Slem (1996) found little evidence that self-managing teams performed better or that employees preferred to work on them. The idea of self-management is attractive to many employees in theory, but so is having a good leader to manage, teach, and reward efforts. The lack of a single best approach to leadership should not be too surprising. When the task is complex and the team's goals are unclear, a strong leader is needed to provide clear direction. When the task is relatively routine, the need for a leader is greatly diminished. The more experience people have in performing the task and working as a team, the better able they are to become self-managing.

5. Application: The Functional Approach to Leading Teams

The functional approach to leadership focuses on how the leader can help the team to work effectively (Hackman & Walton, 1986; Zaccaro & Marks, 1999). From the functional perspective, leadership is a form of social problem solving. The leader monitors the situation, diagnoses problems affecting the team, and implements solutions to these problems. Advocates of the functional approach have identified three areas of focus for team leaders: direction, structure, and external relationships.

One of the primary roles of the team leader is to set the direction for the team. The leader has the important internal role of motivating members and helping them achieve their own goals as well as the team's goals. Establishing a clear and engaging direction for the team is a crucial part of motivating team performance.

The leader has to create a situation that will enable successful performance. Its components include a facilitative group structure, a supportive organizational context, and the availability of expert coaching. A facilitative group structure contains tasks that are engaging, a group whose members have the skills to complete the task, and group norms that encourage effective performance. A supportive context provides the team with necessary information and rewards team excellence. The team leader also is responsible for coaching and facilitating the team's internal group process.

The third leader role is oriented toward the team's external relations. The leader links the team to the organization and buffers the team from interference from the organization. The leader has a public relations job to perform, making sure the team has the resources and support it needs from the organization.

How the leader should intervene to support or coach a team depends on its stage of development. These stages change the readiness of the team to accept and use different types of interventions from the team leader (Hackman & Wageman, 2005). At the beginning of a project, the team needs to become oriented toward each other and prepare to work on the task. A coaching intervention that motivates the team by enhancing commitment to both team and task is appropriate. During the midpoint transition period, strategy-oriented coaching that helps the team analyze and improve operations is required. When most of the team's work has been completed, educational coaching may be used to help the team learn from the work experience and enable members to use these lessons in future team activities.

This chapter concludes with a bit of advice for team leaders. Research shows that leaders who actively listen to team members and incorporate their ideas into the teams' decisions help improve both members' evaluations of their teams and the quality of team decisions (Cohen & Bailey, 1997). Problem leaders tend to micromanage their teams, engage in autocratic decision-making, and be overconfident of their own skills (McIntyre & Salas, 1995). This pattern of leadership reduces respect for such leaders, and prevents constructive feedback from improving their behavior.

Summary

Leadership can be centralized in one person or distributed among various roles. Teams vary in types of leadership, selection of leaders, and delegation of leader powers. Teams rarely exist without leaders, since leaders emerge through a team's interactions. Leaders may be designated by their organizations, or teams may select their own leaders and be self-managing.

There are four main approaches to studying leadership. The trait or personality approach defines the personality characteristics of successful leaders. The behavioral approach examines the value of different behavioral styles such as task orientation versus social orientation. The situational approach identifies the factors that make leaders important (change) or less important (mature teams). The contingency approach links traits and behaviors to the situations to which they best apply. Each of these approaches has different implications for the way leaders should be selected and trained.

One of the most important leadership theories for teams is situational leadership theory, which defines four styles of leadership: directing, coaching, supporting, and delegating. The leader should select the style to use based on the readiness level of the group. In addition, the leader should use an appropriate style to promote team development.

Self-managing teams shift authority and responsibility for teams from management to the team members. The chief examples of self-managing teams at work are factory and service workers, where employees are cross-trained and taught teamwork skills. Developing self-managing teams among professionals and managers can be difficult because of the nature of their tasks and the relationships among team members. The use of self-managing teams requires new roles for team leaders.

The functional approach to leadership provides advice about what factors a leader should focus on to improve the team's functioning. The primary role of the team leader is to provide direction for the team. The leader also is responsible for establishing the structure that supports the team's work. Finally, the leader manages the external relations of the team.

TEAM LEADER'S CHALLENGE—10

You are an attorney in a law office with several other attorneys, paralegal assistants, clerical staff, and an office technician. Although leadership is shared among the attorneys, you are the leader for the staff meetings and for office management issues. Decisions about how the office operates are made in weekly meetings with the entire staff. The office technician has informed you that problems are increasing with the office computer system and it is time to make a major change. This technological decision will affect the work of everyone in the office.

The office functions well and people have good working relationships. Although they do not welcome learning a new computer system, many people in the office recognize that a change in technology is needed. For most office decisions, the staff discusses issues and makes a group decision. However, there are many technical aspects to the computer system decision, and you are uncertain whether everyone should be involved in this decision.

As the leader, how should you make the decision about the new computer system?

Are the team members capable of making this decision, or is this a time when more authoritative leadership is important?

What leadership style is best here? Why?

ACTIVITY: OBSERVING THE LEADER'S BEHAVIOR

Objective. Situational leadership theory defines four types of leader behavior: directing, coaching, supporting, and delegating. The most useful behavior depends on the readiness level of the group. Group readiness relates to the skills, experience, and responsibility level of the group.

Activity. Select a variety of situations to observe and analyze the behavior of leaders. These situations can come from teams or organizations you belong to, videos of leaders interacting with groups, business case studies, or the Team Leader Challenges in this book. Use Activity Worksheet 10.1 to classify the leader's behavior using the types from situational leadership theory, rate the readiness level of the group, and analyze the match between these factors.

ACTIVITY WORKSHEET 10.1

Rating the Leader's Behavior

Which of the following styles best describes the leader's behavior?

_____ Directing (high task & low relationship)

_____ Coaching (high task & high relationship)

_____ Supporting (low task & high relationship)

_____ Delegating (low task & low relationship)

Overall, how would you rate the readiness level of the team?

_____ Low   _____ Medium   _____ High

How well did the leader's behavior match with the group's readiness level?

Leadership Style			
Directing	Coaching	Supporting	Delegating
Low	Medium		High
Group Readiness			

Analysis. According to situational leadership theory, does the style of behavior used by the leader match the readiness level of the group? Was the style of leadership used effective? How should the leader behave to be more effective?

Discussion. What are the implications of using a leadership style that is too controlling or task oriented? What are the implications of using a leadership style to give subordinates too much freedom and responsibility? How will group members respond to these styles of leadership?

11

Problem Solving

Group problem solving has been studied using three different approaches: how groups go about solving their problems, what types of behavior contribute to effective problem solving, and what techniques can be used to improve group problem solving. Teams should base their problem-solving approaches on a rational model of the process that includes six stages: problem definition, evaluation of the problem, generating alternatives, selecting a solution, implementation, and evaluation of the results. In practice, however, this rational approach is rarely followed, and teams often find themselves developing solutions before they understand the problems.

At each stage of the problem-solving process, teams can use a number of techniques to improve their problem-solving abilities. Using these techniques helps teams to be more effective problem solvers.

LEARNING OBJECTIVES

1. How do groups typically solve problems?

2. What factors help improve a group's ability to solve problems?

3. What factors disrupt a group's ability to solve problems?

4. Understand the main steps in the rational approach to problem solving.

5. How do the characteristics of the problem, group, and environment affect the way a group analyzes a problem?

6. What is the value of using a structured approach to generating and evaluating alternatives?

7. What factors affect the implementation of a solution?

8. Why should problem-solving teams use structured techniques to analyze and solve problems?

9. Understand some of the techniques that teams can use to help in their problem-solving efforts.

1. Approaches to Problem Solving

A problem is a dilemma with no apparent way out, an undesirable situation without a solution, a question that cannot currently be answered, the difference between the current situation and a desired state, or a situation group members must manage effectively (Pokras, 1995). The problem can come from the environment or arise from the group. Problems often first surface for a group as symptoms that cause undesirable effects.

In a work environment, a problem for many teams is simply how to complete their tasks or assignments. A team's assignment contains two primary problems: (a) determining the nature of the assignments and how to complete them and (b) managing problems and obstacles encountered when performing them. These obstacles may be technical issues, conflicting viewpoints, or interpersonal conflicts.

The perfect way to solve a problem is to define it and then decide how to solve it. This may seem obvious, but the biggest problem teams have is generating solutions without first understanding the problem. Defining and evaluating the problem is the most difficult step to perform.

The first step in problem solving is to discuss and document individual views until everyone agrees on the nature of the problem (Pokras, 1995). Teams are often given ill-defined problems and undeveloped criteria for evaluating them. Teams need to challenge the definitions of the problems, searching for their root causes. They also need to define what successful resolutions would look like in order to evaluate alternative solutions. The result should be agreement on the issues that need resolution and clear statements of the problem.

Teams may rush through the problem definition stage, only to find that they have to return to it during the solution or implementation stage. This is a time-consuming approach to problem solving. Understanding as much as possible about a problem at the beginning can reduce the overall time spent on a project.

Another common problem is ignoring the final stage: evaluating the solution. Often teams are created to solve problems but are not responsible for implementation or evaluation. Evaluation is ignored because no one wants to present negative information to superiors. Rather than learning from mistakes made, the mistakes are hidden from the team and organization, and as a result they are often repeated because of lack of feedback.

There are three approaches to group problem solving: descriptive, which examines how groups solve problems; functional, which identifies the behaviors of effective problem solving; and prescriptive, which recommends techniques and approaches to improve group problem solving (Beebe & Masterson, 1994).

Descriptive Approach: How Groups Solve Problems

The descriptive approach examines how groups solve problems. Researchers focus on different aspects of the group process in order to understand the problem-solving process. These different perspectives offer alternative ways of understanding the process.

One perspective using the descriptive approach is to identify the stages a group goes through during problem solving (Beebe & Masterson, 1994). This approach is similar to the stages of group development discussed in Chapter 3. The four stages a group uses when solving a problem are forming, storming, norming, and performing.

In the forming stage, the group examines the problem and tries to better understand the issues related to it. The storming stage is a time of conflict, where different definitions of the problem and preliminary solutions are discussed. Often, the group jumps ahead to arguing about solutions before it has reached agreement on the problem, so it must return to the problem definition stage to resolve this conflict. In the norming stage, the group develops methods for analyzing the problem, generating alternatives, and selecting a solution. The establishment of these methods and other norms about how to operate helps the group work together effectively. In the performing stage, these methods are used to solve the problem and develop plans to implement the solution.

Rather than going through problem-solving stages, many groups start the problem-solving process by generating solutions. Groups generate alternatives and select solutions in a variety of ways. These strategies include selecting a solution at random, voting for the best solution, taking turns suggesting each member's favorite solution, trying to demonstrate that a solution is correct, or inventing new and novel solutions (Laughlin & Hollingshead, 1995). Once a solution becomes the focus, the group analyzes it to determine whether it is correct, or at least better than the proposed alternatives. If the majority of members believe that it is, the solution is accepted. If they do not, a new solution is generated by one of the preceding techniques.

2. Functional Approach: Advice on Improving Group Problem Solving

The functional approach tries to improve a group's ability to solve problems by understanding the factors relating to effective problem solving and the factors that disrupt group problem solving.

Factors That Improve Group Problem Solving

An effective group should include intelligent problem solvers or vigilant critical thinkers. The group should analyze the problem, develop alternatives, and select the best solution. The problem-solving process should be relatively free of social, emotional, and political factors that disrupt a rational approach. The following are characteristics of effective group problem solvers (Beebe & Masterson, 1994; Janis & Mann, 1977):

- Skilled problem solvers view problems from a variety of viewpoints to better understand the problem.
- Rather than relying on its own opinions, an effective group gathers data and researches a problem before making a decision.
- A successful group considers a variety of options or alternatives before selecting a particular solution.
- An effective group manages both the task and relational aspects of problem solving. It does not let a problem damage the group's ability to function effectively in other areas.
- A successful group's discussion is focused on the problem. Too often, groups have difficulty staying focused on the issues, especially when there are conflicts.
- An effective group listens to minority opinions. Often the solution to a problem lies in the knowledge of a group member but is ignored because the group focuses on the opinions of the majority.
- Skilled problem solvers test alternative solutions relative to established criteria. The group defines what criteria a good solution must meet and uses those criteria when examining alternatives.

Factors That Hurt Group Problem Solving

Project teams often jump quickly to the solution stage without adequately defining the problems (Hackman & Morris, 1975). The teams do not discuss their problem-solving strategies or develop plans to follow. Typically, they try to apply solutions that have worked in the past. When teams spend time following a structured approach to problem solving, their decisions are better and members are more satisfied with the problem-solving process.

A group may not follow a structured approach to problem solving because of constraints on the process, such as limited time, money, and information. Because of these constraints, groups often seek "satisficing" solutions rather than optimal solutions (Simon, 1979). Perfection is expensive and time-consuming. Collecting all relevant information needed to solve problems may take longer than the time or resources available to groups. In most cases, groups try to find acceptable solutions (those that meet their basic needs) given the constraints of the situation.

In addition, it often is difficult to determine the best solution. There are trade-offs, such as cost versus effectiveness of the solution. Solutions differ according to their probabilities of success, the availability of resources for their implementation, and the politics of implementing them. These trade-offs do not have correct answers; they rely on the judgment of the group members. This limits a group's ability to objectively select the best solution.

As discussed in Chapters 6 and 9, communication problems may interfere with a group's ability to analyze and solve problems. During a group discussion, more time is spent on reviewing shared information than on discussing specialized information that might be pertinent to a solution (Stasser, 1992). Although the group's discussion should be focused on the problem, group discussions can get sidetracked and disrupted in many ways (DiSalvo, Nikkel, & Monroe, 1989). Ideally, a group would spend more of its time sharing information, planning, and critically evaluating ideas than on discussing non-task-related issues. The group's ability to solve problems would be increased, but groups often fail to do this (Jehn & Shaw, 1997).

A group's problem-solving process can be disrupted by a number of non-task-related factors. Group members may support a position because of their desire to reduce uncertainty or avoid social conflict. Politics may encourage members to support alternative solutions out of loyalty to their creators or as payback for past political support. Competition and power differences may disrupt a group's ability to solve problems (Johnson & Johnson, 1997). Competition in the group may encourage political advocacy rather than a search for the best alternative. Groups are better able to solve problems when power is relatively equal among group members. This encourages more open communication and critical evaluation. Groups are better at problem solving when power is based on competence or knowledge, rather than on formal authority.

3. Prescriptive Approach: Rational Problem-Solving Model

The functional approach illustrated what can go right (and wrong) with the group problem-solving process. The prescriptive approach presents a strategy that encourages groups to solve problems more effectively. This approach is based on the assumptions that (a) group members should use rational problem-solving strategies, and (b) using a structured approach will lead to a better solution. The value of formal structured approaches to problem solving varies depending on the type of problem. The more unstructured and complex the problem, the more helpful using a structured approach will be to the group (Van Gundy, 1981).

An outline of the prescriptive approach is presented in Figure 11.1, which shows the main steps in a formal, rational, problem-solving model.

Problem Recognition, Definition, and Analysis

Problem recognition, definition, and analysis are key processes in effective problem solving. However, groups often rush through these stages of the problem-solving process. In their desire to develop solutions quickly, they focus on the symptoms of the problem rather than trying to understand the real causes of the problem (Pokras, 1995).

Unfortunately, even when a group takes the time to identify and analyze a problem, the problem and its causes may be misinterpreted. Many things can go wrong in the problem analysis process. The ability to successfully identify and analyze a problem depends on the characteristics of the problem, the group, and the environment (Moreland & Levine, 1992).

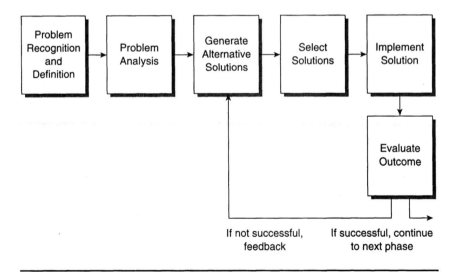

Figure 11.1   Rational Problem-Solving Approach

SOURCES: Adapted from Dewey, J. (1910). *How we think*. New York: Heath. And Van Gundy, A. (1981). *Techniques of structured problem solving*. New York: Van Nostrand Reinhold.

Problems vary in their levels of severity, familiarity, and complexity. The more severe a problem is, the more likely it is to be identified as a problem. Acute problems with identifiable onsets and impacts are often recognized and addressed, whereas chronic problems that are less visible are often ignored. Problems that are familiar are more easily recognized. Novel problems are more difficult to interpret, and groups may assume they are unique, one-time events that will go away by themselves. Complex problems are difficult to analyze and interpret. It is common for a group to select only part of a complex problem to analyze and solve, so as to simplify the situation (although this might not be an effective way in which to resolve the problem).

Groups vary in their levels of desire and ability to identify problems. Group norms have a strong effect on problem identification. Groups with norms supporting communication and positive attitudes toward conflict are more likely to identify and discuss problems. Groups vary in how open they are to the environment. Closed groups that are internally focused are less likely to be aware of problems in the environment. Open groups monitor what is happening in the environment; they are better able to prepare for problems in the future because they have identified the issues beforehand.

Group performance affects the problem identification process. A group that is performing successfully will sometimes ignore problems. From their perspective, the problems cannot be very important given that the group is currently successful. Unsuccessful groups also have a tendency to ignore problems. These groups must focus on their main performance problems, and as a result they are less likely to see other problems. The notion of continuous improvement is a teamwork concept designed to help deal with this issue. In continuous improvement, teams assume that part of their function is to improve operations. In essence, successful and unsuccessful teams are required to identify problems and work to solve them on an ongoing basis.

Characteristics of the environment also affect a group's ability to identify and analyze problems. Many modern environments (e.g., political, business, technological) have substantial levels of change and uncertainty. The rapidity of change creates a need to stay alert and prepare for future problems, while the level of uncertainty makes this more difficult to do. Groups vary in their relations to the outside environment. For example, some work teams are required to accept the definitions of problems given by their organizations, while other teams are open to information about potential problems from outside sources (e.g., customers, suppliers, the public).

Once a group identifies a problem, it may not decide to solve it (Moreland & Levine, 1992). There are other alternatives. The group may decide to deny or distort the problem, thus justifying their choice to ignore it. The group may decide to hide from the problem, given that problems sometimes go away by themselves. If the problem is difficult for the group to understand (due to novelty or complexity), it may decide to just monitor the problem for the time being. Working collaboratively to solve a problem requires identification, belief that the problem is solvable, and motivation to solve it. These are the necessary conditions for the first two stages of the rational problem-solving process.

Generating Alternatives and Selecting a Solution

Finding an effective solution depends on developing high-quality alternatives (Zander, 1994). The ability of a group to accomplish this is related to the knowledge and skills of group members. However, it also depends on the group's climate and processes. The climate of an effective group encourages open discussion of ideas, where minority ideas are heard and taken seriously by the majority.

Groups sometimes use creativity and other structured techniques to generate alternative solutions to problems. Techniques such as brainstorming and the nominal group technique (discussed in Chapter 12) are used to generate alternatives. An important value of these techniques is that participation by all group members is encouraged. However, these participation techniques are useful only if the group is willing to give divergent ideas a fair evaluation. Too often, conformity pressure leads groups to adopt solutions used in the past because of majority support.

After generating alternatives, groups must consider how to determine the best solution. Groups should consider the positive and negative effects of each alternative. The ability to implement the solutions must be considered. This involves the ability of groups to enact the solutions and understanding how outside groups will respond to the solutions.

Any good solution meets three criteria: (a) It is a prudent agreement that balances the needs of various group members; (b) it is an efficient problem-solving approach that does not consume too much time and resources; and (c) it is a process that fosters group harmony (Fisher, Ury, & Patton, 1991). Once a set of alternatives has been developed, the group should not argue about the merits of each solution; to do so encourages a conflict based on positions. Instead, the group should develop ways of evaluating the benefits and costs of the alternatives. The focus should be on analyzing the alternatives to aid selection, rather than on the politics of getting an individual position adopted. Proper focus leads to a final solution containing elements from multiple alternatives.

Sometimes none of the available alternative solutions is appealing, in which case the group selects the least objectionable proposal. This often leads to rationalizing among group members to bolster their belief that the decision is acceptable. Groups may overemphasize the positive attributes of a selected solution and deny its negative aspects in order to justify their choice (Janis & Mann, 1977).

After the group has made its decision, it may want to hold a "second chance" meeting to review the decision. Even when the group decides by consensus, it is useful to have a second-chance meeting to air concerns about the decision. The meeting helps prevent factors such as groupthink and pressure to conform from inappropriately influencing the decision.

Implementation and Evaluation

A solution is not a good one unless it is implemented. This requires commitment from a group to support and enact its solution. As mentioned in Chapters 8 and 9, one of the benefits of group decision making is that participating in the decision process creates a sense of commitment to it.

A problem-solving group is obliged to think about implementation issues when making a decision (Zander, 1994). It is not useful to agree on a solution that cannot be implemented. This means that the group should plan how the solution will be implemented, including consideration of the people, time, and resources needed for implementation. It may be useful to bring the people who will be affected by the planned solution into the decision-making process to encourage their acceptance of the solution.

Evaluation requires examining how the solution was implemented and what the effects were (sometimes called process evaluation versus outcome evaluation). This requires that the group provide a definition of a successful outcome, something it should have done during the problem identification stage.

Sometimes, even when the solution resolves the problem, the undesirable situation does not change significantly. This happens when a group solves only part of a larger problem and the rest of the problem comes to the foreground. By taking a larger perspective on the problem, the group may be able to determine the more critical parts of the problem that should be solved. The evaluation stage provides information for future problem identification and solving.

4. Problem-Solving Teams

Problem-solving teams are typically established for brief periods to solve specific organizational problems or encourage organizational improvements (Fiore & Schooler, 2004). These teams work on a variety of issues such as quality, process improvement, re-engineering, and organizational development. Problem-solving teams may be composed of people from different organizational levels, from production and service employees to professionals and managers, and from different parts of an organization. Consequently, team members often do not know one another's skills or areas of expertise and may have communication problems because of language and background differences. Because of these characteristics, problem-solving teams often rely on facilitators and the use of structured problem-solving techniques.

Teams must have a shared conceptualization of a problem to solve it. A team cannot coordinate its problem-solving efforts without this shared mental model. In problem solving, the mental model includes the nature of the problem, roles and skills of team members, and the mutual awareness of team members. A shared understanding of a problem ensures that all team members are solving the same problem.

A problem-solving team may use an engineering problem-solving technique called process mapping (Fiore & Schooler, 2004). Process mapping works as a problem-solving tool because it leads to the construction of a shared mental model for the team. The team develops a process map of how the situation currently operates (an "as is" map) that defines the parts of a process and the linkages among the parts. The team then develops a "should be" map that describes how the process should operate. These maps are then used to analyze the organization's operations and develop recommendations for improvement.

The value of process mapping is that it facilitates team communication regarding the problem definition, which improves later problem solving. In jointly developing the process map, the team arrives at a shared understanding of the problem. This overcomes the tendency of teams to skip to the solution stages of a problem. It also creates an environment where diverse team members can share their knowledge about the problem.

As team members engage in process mapping, the unique knowledge of each team member is made explicit. The team becomes aware of both the unique and common knowledge it shares. It is forced to negotiate its understanding of the issues related to the problem. Process mapping creates an external representation of a shared problem that facilitates the team's ability to work together to solve the problem. It forces the team to acknowledge deficiencies (the problems in the "as is" map) before attempting to develop solutions.

Process mapping is one of many problem-solving techniques used by these teams (Katzenbach & Smith, 2001). The value of such techniques is that they provide a structure for communications and focus the team on clearly defining the problem before developing solutions. This type of structure is especially important for temporary problem-solving teams because of the sometimes limited English language skills of global production and service workers and the communication jargon of diverse professional teams.

Research demonstrates that structured approaches help these teams to make better decisions, increase members' satisfaction with solutions, and increase commitment to implementation (Pavit, 1993). These problem-solving approaches are effective because they promote more equitable participation in decisions, reduce the negative impact of unequal status, and increase the likelihood that the ideas of low-status employees will be considered.

5. Application: Problem-Solving Techniques for Teams

Teams can choose several useful techniques to help them at each stage of the problem-solving process. These techniques structure the group process and enable the team to better focus on the problem. Four of these problem-solving techniques are presented in Figure 11.2. *Symptom identification* is a technique to help in the problem analysis stage. The *criteria matrix* is used to assist in selecting a solution. *Action plans* improve the implementation of a solution. *Force field analysis* can be used in many stages of the problem-solving process. The following sections examine these and other techniques in more detail.

Symptom Identification

Problem solving begins by recognizing that a problem exists, and that most of the real problem lies hidden. Typically, the first encounter with a problem is only with its symptoms. The team must then find and agree on the fundamental sources of the problem. It should separate the symptoms (which are effects) from the causes. Before using the tools in this approach, team members investigate the problem by gathering more information about it. With this new information, the team can analyze the cause of the problem.

There are several tools that may be useful at this stage (Pokras, 1995). Symptom identification is a technique that uses a simple form to tabulate all aspects of a problem. In force field analysis, the team analyzes the driving and restraining forces that affect a problem. In charting unknowns, team members discuss what they do not know about the problem, which generates hidden facts, questions, and new places to look for information. In repetitive "why" analysis, the team leader states the problem and asks, "Which was caused by what?" This question is repeated several times to examine underlying causes of a problem.

The symptom identification technique uses a chart to display what is known about a problem (Figure 11.2). The chart is a simple form used to tabulate all aspects of a problem. There are two types of facts. Hard facts are

Problem Analysis Stage:
  Symptom Identification Technique

Symptoms	Facts	Opinions

Selection of a Solution Stage:
  Criteria Matrix

Alternative Solutions	Evaluation Criteria			

Problem Analysis Stage:
  Action Plan

Action	Responsible Person	Expected Result	Completion Date

Multiple Problem-Solving Stages:
  Force Field Analysis

Driving Forces What do you want?	Restraining Forces What prevents you from getting it?

Figure 11.2   Problem-Solving Techniques

SOURCE: From Pokras, S. (1995). *Team problem solving*. Menlo Park, CA: CRISP Publications.

based on data and information; soft facts are based on impressions, opinions, and feelings. Team members list all aspects of the problem they are aware of, then note whether there are hard or soft facts that support their beliefs about the problem.

Criteria Matrix

Techniques to generate alternatives are presented in Chapter 12. Once the team has generated alternative solutions, a selection process is required to review and evaluate them. If the group has done a good job generating alternatives, they should have a number of options from which to choose. If the team has used creativity techniques such as brainstorming, there may be many unworkable ideas. Because some approaches obviously are not going to work, they should be eliminated from further analysis. Then the team should review the options and look for ways of combining solutions. After this, the team can develop a criteria matrix to evaluate the alternatives objectively.

A criteria matrix is a system used to rate alternatives (Pokras, 1995). The first step is to decide what criteria will be used to rate the alternatives (or what standards are to be used to evaluate them). These criteria are used to create a chart for evaluating the alternatives (Figure 11.2).

What evaluation criteria should be used? Many options exist, including ease of implementation, effectiveness, expense, and quality. A team may want to use a rating scale for its analysis (e.g., 0 = not acceptable, 1 = somewhat acceptable, 2 = acceptable). It is important to not merely select the alternative with the highest score, given that not all evaluation criteria are of equal importance. The criteria matrix allows the team to analyze and discuss the relative merits of the alternatives in a structured manner.

Action Plans

The implementation stage focuses on generating action plans, considering contingency plans, and managing the project on the basis of these plans. An action plan is a practical guide to translating the solution into reality—a step-by-step road map, if possible (Pokras, 1995). It emphasizes the timing of various parts and assigning responsibility for actions. The plan also should establish standards to evaluate successful performance (Figure 11.2).

Events rarely go as planned. The team should establish a monitoring and feedback system to ensure that team members are aware of the progress made. Larger action items should be broken down into stages and monitored. Feedback to the team on progress with individual assignments should be a regular part of team meetings.

Force Field Analysis

Force field analysis is an approach to understanding the factors that affect any change program (Lewin, 1951). It examines the relation between the driving and restraining forces for change (see Figure 11.2). The driving forces are what the team wants to achieve and the factors that minimize the problem. The restraining forces are the obstacles that prevent success and the factors that contribute to the problem. This approach can be used at many stages of the problem-solving process, but it is especially valuable in examining implementation issues.

When implementing a solution to a problem, teams want to increase the driving forces that encourage the change and reduce the restraining forces that prevent the change from occurring. Teams often focus on the driving forces that are promoting the change. However, most unsuccessful change efforts are due to the restraining forces (Levi & Lawn, 1993). Reducing the power of the restraining forces is a necessary precondition for change.

Force field analysis provides a method for teams to study their problem-solving activities. Using Lewin's action research model, teams use group discussions and surveys to identify the driving and restraining forces affecting any proposed change. The team uses this information to decide on strategies to support the change effort. A cycle of generation, survey, and application of results may be repeated during the stages of the problem-solving process.

Levi and Lawn (1993) used this approach to analyze the driving and restraining forces that affected project teams developing new products. The project teams were driven by interest in new technology and an organizational culture that encouraged innovation. However, the success of producing and marketing new products was restrained by technical problems in manufacturing and financial issues. Understanding these forces encouraged the project teams to include members from manufacturing and marketing in the design teams to address these problems.

Summary

Problem solving requires that a team analyze the nature of the problem, then develop and implement a solution. Unfortunately, many things can go wrong during these two steps. The study of group problem solving uses descriptive, functional, and prescriptive approaches to understand and improve the problem-solving process.

The descriptive approach looks at how a group solves a problem. The problem-solving process goes through developmental stages similar to overall stages of group development. Solutions are often generated in a rather haphazard fashion that seems more political than logical.

The functional approach provides advice on how to improve the group problem-solving process. An effective group views problems from multiple perspectives, analyzes a variety of alternatives using established criteria, and manages the group process to ensure that all members may participate. The group's ability to solve problems may be hurt by rushing to the solution stage, by constraints limiting the amount of analysis, by confusion about evaluation criteria, and by social factors that disrupt the group process.

The prescriptive approach to problem solving includes a series of structured stages. The problem identification and analysis stage is affected by the severity and complexity of the problem, group norms about discussing problems, and the amount of uncertainty in the environment. The process of developing and selecting alternative solutions is improved by creativity techniques to generate alternatives and by analysis techniques to examine alternatives in a systematic manner. Implementing solutions requires planning and an evaluation system to provide feedback on the process.

Organizations use temporary problem-solving teams to deal with a variety of issues and to encourage improvement. These teams function more effectively if they use structured techniques, such as process mapping.

The team may use a variety of techniques to improve its problem-solving skills. Symptom identification techniques help clarify what is known about a problem. A criteria matrix is used to evaluate alternative solutions. Action plans create a map to guide implementation. Force field analysis may be used at several stages to evaluate alternatives and implementation programs.

TEAM LEADER'S CHALLENGE—11

Your organization uses improvement teams composed of professionals and managers throughout the organization to solve important organizational problems. Team membership is highly valued because participation provides good visibility to upper management. Consequently, team members are highly motivated to perform. You have been selected to lead the next team. To prepare for the role, you have been discussing problems with former team leaders.

The last improvement team got off to a fast start. At the first meeting, they diagnosed the problem and started generating solutions. They quickly focused on a preferred alternative and began developing an implementation program. After several months of work, the team presented their proposal to top management. However, when they started implementing the proposal, serious problems became apparent and the project was scrapped.

How can you avoid the problem of the previous project team?

What problem-solving approaches should you use?

How can you prevent the team from wasting time on a proposal that doesn't really solve the problem?

ACTIVITY: USING PROBLEM-SOLVING TECHNIQUES

Objective. Problem-solving is improved when a team follows a structured approach. The team should analyze the problem thoroughly before developing alternatives. It should develop a set of alternatives and then use evaluation criteria to help select a solution. Force field analysis can be used to understand the issues related to implementing a solution.

Activity. Have the team follow a structured approach to problem solving. The team can be given either an organizational problem or a social problem to solve. For example, develop a program to reduce cigarette smoking, increase the use of seat belts, or encourage the use of condoms. After a problem has been selected, the team should use the symptom identification chart to understand the problem, develop several alternative solutions, use the criteria matrix to analyze the alternatives, select an alternative solution, and use force field analysis to understand the issues that will affect implementation of the solution.

Step 1: Analyze the symptoms of the problem using the symptom identification chart (Activity Worksheet 11.1). This symptom analysis should help you to understand the different causes of the problem.

ACTIVITY WORKSHEET 11.1

Symptom Identification Chart

Symptoms	Facts & Data	Opinions & Feelings

SOURCE: From Pokras, S. (1995). *Team problem solving.* Menlo Park, CA: CRISP Publications.

Step 2: Generate alternative solutions to the problem.

Step 3: Develop criteria that can be used to evaluate the alternative solutions. Criteria may relate to issues such as quality, cost, and acceptability.

Step 4: Use the criteria matrix (Activity Worksheet 11.2) to analyze the alternative solutions. Rate each solution on the criteria that your team has developed. Then select a preferred solution.

ACTIVITY WORKSHEET 11.2

Criteria Matrix

Alternative Solutions	Evaluation Criteria			

SOURCE: From Pokras, S. (1995). *Team problem solving*. Menlo Park, CA: CRISP Publications.

Step 5: Evaluate the implementation of your solution using force field analysis (Activity Worksheet 11.3).

ACTIVITY WORKSHEET 11.3

Force Field Analysis

Driving Forces What do you want?	Restraining Forces What prevents you from getting it?

SOURCE: From Pokras, S. (1995). *Team problem solving*. Menlo Park, CA: CRISP Publications.

Analysis. Did the team members find the use of the structured problem-solving approach helpful? What aspects of it did they like or dislike? Did it improve the quality of the solution?

Discussion. What are the advantages and disadvantages with using a structured approach to problem solving?

12

Creativity

Developing creative solutions to problems is an important concern for teams, and teams can use a number of techniques to stimulate group creativity. However, the dynamics of groups tend to limit creativity because of cognitive, social, and organizational problems. The solution to promoting creativity in teams requires approaches that combine the benefits of individual and group creativity. Organizations can help to encourage creativity by providing supportive organizational climates.

Learning Objectives

1. Understand the different ways of defining creativity.

2. What are the psychological factors that help and hurt individual creativity?

3. Why do groups have problems in developing creative ideas?

4. What factors improve groups' ability to be creative?

5. Why do organizations have mixed views of the value of creativity?

6. What are some of the organizational factors that affect creativity?

7. Understand how to use brainstorming, the nominal group technique, and brainwriting to improve group creativity.

8. What are the advantages of using multiple group sessions to encourage creativity?

1. Creativity and Its Characteristics

Creativity can be defined from the standpoint of a person, a process, or a product (Amabile, 1996). From this perspective, creativity is something that creative people do, some things or ideas are creative, and it is a process that produces creative things or ideas.

Most research examines creativity from the person perspective, trying to find out who is creative by using personality tests or other psychological measures. However, there are problems with the person-oriented approach. Creativity varies in degree; it is not simply a personality trait that some people have and others do not. Creativity skills can be enhanced through training for both individuals and groups. Creativity relates to the match between the person and the application. In other words, people are not creative in all areas. Talent, learned skills, and situational factors affect creativity. Creative talent alone is insufficient.

From a teamwork perspective, it is more useful to define creativity in terms of its products and processes than in terms of personality, because this shifts the focus from the individual to the group. What is a creative product or idea? How should a group act to be creative? The answers to both questions show the dual nature of creativity (Table 12.1). It is the search for novel and useful ideas as well as the balancing of divergent and convergent thinking. In the generation stage, people or teams use divergent thinking to develop novel ideas. In the application stage, they use convergent thinking to make those ideas useful.

TABLE 12.1

Aspects of Creativity

	Generation	Application
Product	Novel	Useful
Process	Divergent	Convergent

Something is creative if it is novel and appropriate, acceptable, or useful. Novelty is not sufficient; a creative solution also must be effective. Because of this, creativity requires a combination of divergent and convergent thinking processes (Van Gundy, 1987). Divergent thinking generates potential ideas; convergent thinking analyzes and focuses solutions. These two types of thinking relate to the novelty and usefulness aspects of the definition of creativity.

To encourage divergent thinking, people need to suspend judgment of ideas. Participants should try to generate as many ideas as possible and be receptive to new ideas. Creativity also requires time for incubation, given that creative solutions often occur to people when they are not thinking about the problem. Ideas need to be combined, modified, and played with to encourage creative alternatives.

Taking a systematic approach to selecting a creative solution to a problem encourages convergent thinking. Evaluation criteria are developed and then used to sort out the many ideas generated. All the alternatives are analyzed to avoid premature acceptance. People need to be realistic and not too critical of the ideas of others. The overall goal is to analyze and select the best available alternative.

The creativity of teams relates to individual, group, and organizational factors. Managing the relationships among these factors is the key to promoting creativity within teams. Teams need to have creative members who effectively work together in a supportive organizational context.

2. Individual Creativity

Individual creativity develops from an interaction of personal and situational factors (Amabile, 1996). People are creative when they have domain-relevant skills, creativity-relevant skills, and task motivation.

Domain-relevant skills are the skills, knowledge, and talent people have in a particular application area. People are not creative in all areas; they are creative in areas in which they are skillful. For example, artists are not creative bridge designers, and a creative engineer may not be artistic.

Creativity-relevant skills are the appropriate cognitive styles that encourage creativity and knowledge about creativity techniques. Cognitive styles include the ability to break out of established mental sets, appreciate complexity, suspend judgment, and use broad categories to view issues. Creativity techniques are approaches to helping people play with ideas and view problems from alternative perspectives.

Task motivation includes intrinsic motivation, attitude toward the task, and the impact of environmental factors. Intrinsic motivation is motivation from personal interest or desire rather than from external reward. Motivation is needed to encourage people to apply their creative skills, but some types of motivation are more helpful than others.

Individual creativity can be disrupted by psychological factors. Many of these factors operate by shifting focus away from the task and toward issues external to the task. Individual creativity may be limited by the use of extrinsic rewards, evaluation apprehension, and being stuck in one's paradigms.

In general, intrinsic motivation encourages creativity, whereas extrinsic motivation is a detriment to creativity (Amabile, 1996). Intrinsic motivation relates to the qualities of the task itself. When people are intrinsically motivated, they engage in an activity for its own sake, not to achieve a reward for performing the task. Extrinsic motivation is related not to the task, but to a reward or incentive for performance. Extrinsic motivators can hurt creativity by shifting attention away from the task to receiving the reward.

Extrinsic motivators can hurt creativity in several ways. Being rewarded to perform a task may sometimes reduce the intrinsic motivation for the task (Deci, 1975). A person may enjoy the creativity of painting, but being paid to paint all day may reduce the personal reward of painting. The use of rewards focuses the creator on satisfying the person or organization providing the rewards. Rather than trying to be creative, individuals are trying to satisfy someone else, which may cause them to be more conservative in what they produce.

Rewards do not invariably reduce creativity; they may foster creativity if they signify competence or enable the person to perform more interesting activities. For example, Hewlett-Packard rewards employees who go against management by working on creative ideas even after being told to stop. The rewards acknowledge creativity and give positive feedback to employees. In addition to recognition, employees are given more freedom to work on their own projects. This type of reward encourages others to pursue their own creative ideas.

When someone talks in a group, the listeners are often thinking about what is wrong with what is being said, rather than actually listening to the ideas. This evaluation process discourages creativity (Amabile, 1996). When group members are concerned about appearing stupid, outrageous, or inappropriate, their anxiety limits their creativity.

Evaluation apprehension hurts creativity—especially the ability to generate novel ideas or solutions. It has a more detrimental impact on people with low skills and low self-confidence. An exception to this negative impact is found in work environments in which evaluations focus on providing information or feedback on ideas, and the organizational climate uses positive recognition to reward competence. Here, evaluation may have a positive impact on creativity.

Our internal mental sets or paradigms limit creativity. It is difficult to view a situation from a new perspective. Many famous creative ideas have come from looking at something commonplace from a different perspective. For example, Post-it Notes came from wondering about the value of glue that does not stick very well. Many creative ideas in science come from younger scientists who are not fully indoctrinated into the existing paradigm, or from interdisciplinary scientists who are changing fields. We become locked in an old way of thinking and stuck in our paradigms or routine way of working, so we do not see creative alternatives to our situation.

3. Group Creativity

Creativity is a problem for both individuals and groups. Groups have been shown to be less creative than individuals in some circumstances. Brainstorming, the best-known technique for encouraging group creativity, has been criticized as being ineffective. However, creative groups are sometimes able to overcome these problems.

Problems With Group Creativity

When it comes to creativity, groups face many of the same problems as individuals. When people try to creatively solve a problem as a group, they typically produce fewer ideas than the sum of individuals working alone (Amabile, 1996). Even working alone in the presence of others reduces an individual's creativity. This is especially true if the others observe and evaluate what individuals are doing.

Several group dynamic factors limit creativity (Van Gundy, 1987). Groups may develop negative or critical communication climates that discourage creativity. Interpersonal conflicts in groups may discourage creativity. Groups are more time-consuming, making it no faster or more efficient to use groups. Finally, conformity pressure and domineering members can hurt creativity in the group process.

Cognitive interference and social inhibitors are the main reasons group interaction leads to less creativity than when individuals work alone (Paulus, 2000). Cognitive interference is the disruption of thinking that occurs while one is waiting a turn to speak in a group (also called "production blocking"). During the wait, creative ideas are forgotten, time runs out before the member has a chance to present the idea, and group discussion may drift to irrelevant topics. Social inhibitors relate to anxiety about how others will evaluate one's ideas and social loafing, which is the reduction in motivation caused by one's performance being hidden in the group's output.

Brainstorming

The best-known and most widely used group creativity technique is brainstorming (Osborn, 1957). Brainstorming was designed to deal with the problem of using group discussions for creativity. In group discussions, groups spend too much time evaluating and criticizing ideas and not enough time generating ideas. The four basic rules of brainstorming are: (a) Criticisms are strictly forbidden; (b) free thinking and wild notions are encouraged; (c) numerous ideas are sought; and (d) combining and building on the ideas of others is good. (How to conduct a brainstorming session is presented at the end of this chapter.)

Brainstorming improves creativity relative to unstructured group discussions (Stein, 1975). It can be improved by asking participants to think of ideas alone before the brainstorming session by facilitating the group discussion to make participation more equal, and by ensuring that people do not criticize others' ideas. The primary benefit of brainstorming is separating the generation of ideas from the evaluation of ideas. This reduces criticisms during the discussion of new ideas and encourages the participation of those who would otherwise remain silent.

Research on the effectiveness of brainstorming often shows that it is not superior to individuals working alone (Mullen, Johnson, & Salas, 1991). Brainstorming does not increase the number or quality of creative ideas when compared with the sum of individuals working separately. The chief problem with brainstorming is that group discussions force people to wait their turn (Diehl & Stroebe, 1987). Members in a brainstorming session must take turns speaking, rather than stating ideas when they first pop up. While people are waiting their turns, they are not using their time effectively by developing other creative ideas.

Despite the negative research on brainstorming, it remains a popular technique, especially in business organizations. There are several reasons for this (Parks & Sanna, 1999). First, people in business believe that group interaction stimulates others. This is such a compelling idea that people are unwilling to reject it on the basis of research. Second, people involved in brainstorming sessions believe that brainstorming works; their personal experience supports its benefits. Finally, participating in a brainstorming session may encourage commitment to the final solution.

Recent research on virtual brainstorming (discussed later in Chapter 15) suggests that computer-based forms of brainstorming are more effective than traditional brainstorming in improving group creativity (Dennis & Valacich, 1993). Using communication technology minimizes problems of production blocking and evaluation apprehension. In virtual brainstorming, people are able to review others' ideas and develop and modify their own ideas at their leisure.

Research on virtual brainstorming has shown a number of interesting effects. In computer groups, instead of reducing creativity, group size increases the number of creative ideas. The number of ideas also increases when group members receive feedback on their performance instead of being discouraged by evaluations. The anonymity of the communication medium seems to encourage people to make more comments on others' ideas.

Strengths of Group Creativity

Using groups to develop creative solutions to problems has its benefits. Groups are able to develop more ideas than are single individuals. The social interaction of working in groups can be rewarding. Groups can create supportive environments that encourage creativity. Diverse groups are more likely to develop creative solutions than homogeneous groups. Groups provide support for the implementation of creative ideas.

One key to encouraging group creativity is dissent (Nemeth, 1997). A group with creative conflicts produces more creative ideas. When the group is exposed to contradictory ideas from some members, the thinking of the majority is stimulated and they produce more creative ideas. Dissent stimulates divergent thinking and encourages the group to view an issue from multiple perspectives. It encourages more original, less conventional, thoughts about the issue. These effects occur even when the group does not accept the minority's opinion as correct.

As discussed in the next chapter, diversity in background, training, and perspective increases group creativity (Jackson, 1992). Diversity increases cognitive conflict on issues. Diverse groups generate more ideas, try out more novel ideas, and view issues from multiple perspectives. This effect is one reason why organizations use cross-functional teams that include members with different areas of expertise.

How groups operate can have a positive effect on group creativity. Brainstorming is better for creativity than unstructured group discussion. When trained facilitators run brainstorming sessions, groups generate more creative ideas (Offner, Kramer, & Winter, 1996). This creativity enhancement continues to affect the group after the facilitator leaves because the group learns how to do brainstorming more effectively.

Several methods can improve the brainstorming process to further promote creativity (Paulus, 2000). First, teams should use facilitators to structure the group interaction to avoid disruptive communications and premature evaluations. Group sessions should be followed by individual sessions. Groups should use organizing techniques to reduce the number of alternatives to be evaluated after the idea-generation stage. Finally, groups should be diverse. This diversity must be managed to reduce potential conflict, make members aware of the expertise of others, and focus the discussion on the unique contributions of individual members.

One of the values of group brainstorming is the stimulation of ideas (Paulus, 2000). There is a cognitive benefit to being exposed to other people's ideas because creative ideas often occur through unique associations with other ideas. Exposure to other ideas can help a member break out of limiting cognitive categories. This effect is increased if the group has a diversity of knowledge, experience, and perspectives.

Although group brainstorming may stimulate creative ideas, it may be difficult to demonstrate the benefits of the group interaction (Paulus, 2002). Group interactions limit the ability of individuals to verbalize ideas. It takes time for people to fully process and develop new ideas. The stimulating benefits of group interaction may occur later, when group members have a chance to think about what has occurred. An incubation period is important for both individual and group creativity; this shows the importance of taking breaks during the creativity process and having multiple creativity sessions. Group brainstorming sessions should be followed by individual idea-generation to incorporate the benefits of both individual and group creativity (Paulus, 1998).

Group creativity relates to the individual, the group, and the organization or environment. Are groups more or less creative than individuals? It depends on how the process is managed. The question is really not relevant in many organizational situations. Important problems often require individual and group creativity because the problems are too complex to rely solely on individual creativity.

4. Organizational Environment and Creativity

Organizations benefit from creativity. Businesses want to be innovative to create new products and services to expand their operations. Companies get stuck in old patterns of behavior and look for new ideas to help them break out. The world is a dynamic and turbulent place with a fast rate of change, and organizations must change creatively to survive. To be more creative, they need to hire creative people, use group creativity effectively, and establish organizational climates that promote creativity.

Organizations must rely on group creativity because the problems they face are too complex for individual solution. Often, creative solutions for problems require a multidisciplinary perspective. The development of the first Macintosh computer is a good example. How can one creatively redesign the personal computer, considering all the electronic, manufacturing, artistic, psychological, and human factors involved? It is not a job for an individual; it is a job for a design team. To combine the talents of a variety of fields, organizations need group creativity.

Organizations also must develop creative solutions for problems that cut across organizational boundaries. This too requires a group perspective to fully understand and integrate the issues involved. The use of a group to develop creative solutions for cross-departmental problems also encourages support for the implementation of solutions.

Impact of the Organizational Environment on Creativity

Individual and group creativity can survive only in organizational environments that support it. Although organizations say that they want to encourage creativity, their actions may not support creativity. Organizations want both stability and change, and this contradiction creates problems. There are many things that organizations could do to promote creativity, but there are probably as many obstacles to prevent it from happening.

Organizational leaders say creativity is important and point to famous examples of innovations that have led to organizational renewal and growth. But is this what normally happens? In the business world, creative people are deviants whose ideas are likely to be rejected by organizations. Organizations really tend to value consistent and predictable performance over creativity.

Creativity implies risk. Organizations focus on providing consistency, minimizing error, and reducing risk. This is the inherent conflict with organizational creativity. The problem is not that there are no creative individuals and groups in organizations, but rather that their creativity is not recognized or rewarded.

To foster creativity, organizations must develop climates that support creative people and teams. Organizational climates should promote both the task and social aspects of creativity (Van Gundy, 1987). Climates that support the task aspects of creativity provide the freedom to do things differently, empower people to act on their ideas, encourage active participation, and provide support to those involved in creative tasks. Climates that support the social aspects of creativity allow the open expression of ideas, encourage risk taking, promote acceptance of unusual and novel ideas, and reflect belief and confidence in their employees.

Work environments may either stimulate or provide barriers to creativity. Table 2.2 presents a list of environmental factors that research has shown to affect creativity in the workplace (Amabile, 1996).

TABLE 12.2

Environmental Stimulants and Obstacles to Creativity

Factor	Stimulant to Creativity	Obstacle to Creativity
Freedom	Employees need the freedom to decide what tasks to perform, how to perform them, and control over their work process.	A lack of freedom in the way employees select projects and perform tasks discourages creativity.
Management	Managers need to be good role models, have good technical and communication skills, provide clear directions but use limited controls, and protect teams from negative organizational influences.	A management style that hurts creativity includes unclear direction, poor technical and communication skills, and too much control.
Encouragement	New ideas need to be encouraged, and there should be no threat of evaluation.	A lack of support or apathy toward new approaches reduces the motivation to be creative.
Recognition	Employees should believe that creativity will receive appropriate feedback, recognition, and reward from the organization.	Inappropriate, unfair, and critical evaluations and the use of unrealistic goals limit creativity.
Cooperation	The organizational climate should support cooperation and collaboration, acceptance of new ideas, rewards for innovation, and the allowance of risk taking.	Interpersonal and intergroup competition within an organization or work group disrupts the creative process.
Time	Creativity requires time to explore new ideas rather than rigid schedules.	Too great a workload, or day-to-day crises that redirect focus away from long-term projects, reduce creativity.
Challenge	Tasks that are interesting, important, and not routine encourage creativity.	Organizations that emphasize consistency and do not support risk-taking discourage creativity.

Factor	Stimulant to Creativity	Obstacle to Creativity
Motivation	There should be a sense of urgency from oneself to complete the task because of competition from outside the organization or the desire to do something important.	Organizational factors such as bad reward systems, excessive bureaucracy, and a lack of regard for innovation reduce creativity.

SOURCE: From *Creativity in Context* by Teresa Amabile. Copyright © 1996 by Westview Press, a member of Perseus Books, L.L.C.

In their research on visionary companies such as Hewlett-Packard, IBM, General Electric, 3M, and Disney, Collins and Porras (1994) found these innovative companies have strong core values and socialization programs that instill these values in new employees. The core values strengthen commitment to their companies and increase cohesiveness. This loyalty to the organizations helps employees implement new ideas, but it does not encourage them to be creative. Creativity requires fresh perspectives, a questioning of the way the organizations operate, and deviation from the norms.

These visionary companies have developed a number of mechanisms to promote creativity (Nemeth, 1997). Their organizational norms encourage diversity of opinions and ideas. They provide support for mavericks and individuals with creative ideas, while limiting the fear of failure and providing rewards for risk-taking. In addition, these companies have developed specific programs to encourage creativity. Hewlett-Packard rewards creative defiance of the organization. 3M rewards creative employees with "free time" to work on any project they want. Dupont provides funds to work on ideas not supported by management. Pfizer moved its research and development facilities away from the corporate center to reduce the influence of management on the research staff.

5. Application: Group Creativity Techniques

Developing creative ideas is an important part of a team's work. The tools a team uses to promote creativity can be applied in a variety of ways; creativity techniques, for example, are useful in all stages of problem solving (Van Gundy, 1987). The techniques help clarify objectives, define and analyze problems, generate alternative solutions, and prepare for implementation.

Premature evaluation is the biggest problem limiting team creativity. Team members may want to try out new ideas, but critical comments from other members prevent their consideration. Team members often are not good at supporting one another's ideas, making designated noncritical times vital to creativity. A team should develop rules for openness and safety in presenting ideas; members should practice the technique of building on an idea rather than criticizing it. If they do not like an idea, team members should try going with it by suggesting other related ideas that do not have their objection. Learning this skill is very helpful.

A team often need not decide on an outcome right at that moment. It is better to run a brainstorming session, and wait until the next meeting before selecting an alternative. Waiting allows members to come up with fresh ideas on their own. Such an approach to team creativity captures both individual creativity (so often done alone) and synergistic creativity (arising from group interaction).

Developing creative ideas requires more than just a group session (Figure 12.1). The process begins by developing an open climate that encourages participation. Team members are more likely to develop creative ideas if they have time to prepare for and research the topic. After generating creative ideas, the team selects the best ideas and refines them. Multiple sessions may be necessary to fully develop useful creative ideas.

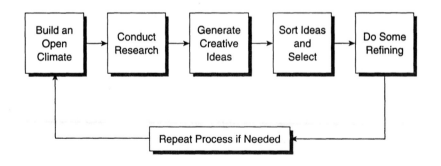

Figure 12.1   Creativity Flowchart

Brainstorming

Brainstorming includes a variety of methods for structuring group creativity sessions. Besides classic brainstorming, alternatives include procedures that force group members to combine the ideas of others, use of pictures to "comment" on ideas, and role playing alternative perspectives such as from cartoon characters. Group facilitators who use brainstorming often have techniques to equalize the level of participation from group members.

To start a brainstorming session, the team leader clearly states the purpose or issues to be discussed and reviews the guidelines for brainstorming (Table 12.3). A distinct period of time (perhaps 10 to 20 minutes) should be set aside for the brainstorming session. During the session, the leader acts primarily as facilitator and recorder. After the ideas have been generated, the leader helps the group prioritize the list into a manageable size for further consideration.

TABLE 12.3

Guidelines for Brainstorming

Question	Announce the question or issue to be addressed.
Toss Out	All team members toss out as many ideas as they can.
Accept	All ideas are accepted, regardless of how practical they are.
Record	All of the ideas are listed for everyone to see.
Prompt	The facilitator asks the main question again to help keep people on track.
No Editing	The facilitator reminds the team that no one is allowed to criticize or evaluate until the process is done.
Build	Everyone should build on one another's ideas, using the ideas as a springboard to new directions.

SOURCE: From Pokras, S. (1995). *Team problem solving.* Menlo Park, CA: CRISP Publications.

During the brainstorming session, all team members should try to suggest as many ideas as possible. Every idea is accepted by the team and written down by the recorder. The leader's job is to keep team members on track by refocusing them on the issue. No one is allowed to criticize ideas; rather, members are encouraged to build on suggestions made by others. It is up to the leader to ensure that no criticisms occur during the brainstorming session.

Nominal Group Technique and Brainwriting

The nominal group technique and brainwriting are similar approaches that combine the benefits of individual and group creativity. As with brainstorming, they separate the idea-generation stage from evaluation. However, in these approaches, individuals generate their ideas in writing rather than in group discussion.

Both techniques start with the same general approach as does brainstorming. A group is brought together, and the facilitator announces the question. In the nominal group technique, each participant spends 10 to 20 minutes writing down ideas. After this stage has been completed, the ideas are listed for all participants to see, and the group is able to ask clarifying questions about the ideas.

Brainwriting has several variations. One approach is to ask members to write down an idea on a sheet of paper and then pass the paper to the person on the right. The next person is required to write a new idea that builds on the previous idea(s). This cycle is repeated until either time is up or the group is exhausted. An alternative is for each group member to write down several ideas on a sheet of paper, throw the papers into a central pool, and pull out another member's paper to write on. Again, group members are to build on the ideas presented in the papers they choose. When the group is finished generating ideas, all the lists are combined for group review.

These approaches are effective alternatives to brainstorming. Because individuals are writing down their ideas, no one must wait a turn to contribute. The techniques are structured to encourage participation from all group members. Typically, they include group discussions at the end so that members have the opportunity to build on one another's ideas.

Selecting a Solution

A problem with creativity activities such as brainstorming is that they generate many possible options and no clear or easy way of selecting the best one. However, after a brainstorming session, it is often easy to prioritize the suggestions and focus on a limited number of options.

One approach to this is multiple voting (Scholtes, 1994). A team reviews the alternatives generated by the brainstorming session and combines items that seem similar. Each team member then selects two to five alternatives that he or she would like to support. When all team members have completed their selections, the votes are tallied and items that received zero to one vote are removed. The alternatives that have been selected are discussed, and the group considers new ways of combining or synthesizing alternatives. These steps are repeated until only a few options remain from which the team can select, at which stage the group may use consensus to select the final alternative.

Multiple-Stage Creativity Approaches

To gain the benefits of group and individual creativity, teams may use a multiple-stage process (Paulus, 1998). This approach uses time as a buffer between group creativity activities. People are often creative at odd moments when they are not thinking about the problem, such as while walking or taking a shower. It is difficult to be creative on command, especially in front of others. Creativity is hard to rush, given that stress and time pressure tend to make individuals and groups more conservative.

Group creativity research unfortunately assumes that groups must be creative on demand. In most organizational contexts, a group works on problems over time. They meet and discuss the problem; then the group tries to be creative; sometime later the group reconvenes to make its selection process. By separating activities in time and allowing group members to enter new ideas over time, the benefits of both individual and group creativity are realized.

Summary

From a group dynamics perspective, creativity typically is defined as a product or process. Creativity leads to the development of what is both novel and useful. The creative process uses divergent and convergent thinking to develop these creative ideas.

Individuals and groups have difficulty being creative on demand. Creativity requires skills in the topic area, creativity skills, and motivation. Extrinsic motivators, evaluation apprehension, and rigid paradigms reduce individual creativity. Groups are often not more creative than individuals working alone. Cognitive and social factors related to the group process limit creativity. Brainstorming is better than unstructured discussions, but has a limited effect on improving group creativity. However, groups can encourage creativity by effectively managing the group process, acknowledging the value of constructive conflict, embodying a diverse membership, and holding multiple sessions.

Although organizational leaders claim they want to encourage creativity, their actions often do not match their words. Creativity is required for organizations to adapt to the changing environment; however, organizations tend to encourage consistency and stability rather than innovation. Organizations can either encourage or discourage creativity through factors such as management orientation, availability of resources, recognition for risk taking, and cooperative climate.

Certain techniques can improve group creativity. Brainstorming structures group discussions to reduce the negative effects of evaluation. The nominal group technique and brainwriting combine individual and group creativity techniques. Selection techniques may be used to reduce the number of alternatives. It is important for groups to use creativity techniques, but it also is important to use multiple group sessions so that creative ideas have time to incubate. When these techniques are used, group creativity is improved.

Team Leader's Challenge—12

You are the leader of a team of writers and artists at an animation studio. This is a great group of highly talented and creative people, but not an easy team to lead. Team members are individualistic, idiosyncratic, and temperamental, to mention just a few of their personality traits. You have been able to organize them into a project team and facilitate their interpersonal issues, and have succeeded in producing a successful short film.

You feel lucky that your team was successful in the last project. It is now time to start over again with a newly composed team. Whether you organize a new team or use the previous team, it is difficult to get a team to work together creatively. What you really want to do is to set up a mechanism to encourage continuous creativity. It is more than just hiring creative people; you want to use teamwork as a way to make creative projects a regular occurrence in the organization.

What are the benefits of and problems with using a team approach to creative work?

How can teamwork be used to encourage creative work?

How can you use teams to ensure creative projects keep flowing from the organization?

Activity: Comparing Different Creativity Techniques

Objective. Creativity may be improved by using one of the creativity techniques presented in this chapter. Teams can try out these techniques to see how well they work.

Activity. Divide into two teams and have each team use a different creativity technique—either brainstorming or brainwriting. Each of these is useful for generating alternative solutions. Spend about 20 minutes using the creativity technique. Then use the multiple voting technique to select the preferred alternative. For a creative challenge, try developing a new advertising slogan for your organization, or write a creative caption for a cartoon in *The New Yorker* magazine.

Brainstorming: To start a brainstorming session, the team leader must clearly state the issues to be discussed and review the guidelines for brainstorming. During the brainstorming session, all team members should try to suggest as many ideas as possible. Every idea is accepted by the team and noted on the recorder's sheet. The leader's job is to keep team members on track by refocusing them on the issue. No one is allowed to criticize ideas; instead, members are encouraged to build on the suggestions made by others. It is up to the leader to make sure no criticisms occur during the brainstorming session.

Brainwriting: Brainwriting starts with the same general approach as brainstorming. Teams are brought together, the leader announces the issue, and team members are told to be open and build on each other's ideas. The difference is that the team's interaction is in writing. Have each person write down several alternative ideas, throw their paper into a central pool, and pull out someone else's paper to add to. Team members are to build on the ideas presented in the lists they receive. When the team is through generating ideas, combine all lists for the team to review.

Selecting a solution using multiple voting: The team reviews the alternatives generated by the creativity session and should combine items that seem similar. Each team member selects two to five alternatives that he or she would like to support. After all team members have completed their selections, tally the votes and discard items that received zero or one vote. Discuss the alternatives that have been selected and look for ways to combine or synthesize them. Repeat these steps until only a few options remain from which the team can select.

Analysis. Which technique generated the most creative solution? What did team members like and dislike about the creativity technique they used? Would they want to use the technique in the future?

Discussion. What are the advantages of and problems with using these two creativity techniques? How can you encourage a team to be more creative?

13

Diversity

Diversity in a group stems from differences in psychological, demographic, and organizational characteristics. Research shows various effects of this diversity, with results depending on how the research is conducted, the type of diversity examined, and the type of tasks performed. In most cases, diversity is a benefit once a team learns how to manage its diversity issues. A group with diverse members performs better on production, problem solving, and creativity tasks.

Diversity in a group can lead to problems caused by misperceptions about others and competition among groups. These problems disrupt group communication and reduce the ability of the group to fully use its resources. Diversity problems are unlikely to go away by themselves, but actions can be taken to help teams improve relations among their members.

Learning Objectives

1. Why is the importance of managing diversity increasing for organizations?

2. What are the different types of diversity?

3. Understand the trait approach versus the expectation approach to explaining diversity.

4. How do cognitive processes, leader behavior, and competition explain the cause of diversity problems?

5. What are the main problems that diversity causes for groups?

6. Understand the differences between the performance of homogeneous and heterogeneous groups.

7. What are the challenges of using cross-functional teams?

8. What are some of the approaches organizations can use to manage diversity problems?

1. The Nature of Diversity

From sociological and organizational perspectives, the topic of diversity is increasing in importance. Diversity has many meanings, which have different impacts on how groups function. This chapter examines these issues and the psychological and sociological causes of diversity problems, the impact of diversity on team functioning, and how organizations can better manage diversity issues.

Why Diversity Is Important Now

Understanding diversity in work teams has become more important during the past two decades, in part because increasing numbers of women and ethnic minorities have been entering the workforce. However, other changes are also making these demographic changes more important (Jackson & Ruderman, 1995). Women and ethnic minorities are now in all levels of the organizational hierarchy; it is no longer the case that they primarily work only in certain types of jobs (Jackson, 1992). In the new organizational environment, diversity occurs in jobs in all areas of organizations.

Age diversity is also increasing in organizations. People are living longer, retirement ages are being extended, and the baby boomers are getting older (Carnevale & Stone, 1995). The relationship between younger and older workers is changing. Younger and older workers are more likely to work together because organizational hierarchies are flatter. New technology has reversed some of the differences between younger and older workers; younger workers may be more skilled with technology and may serve as mentors for older workers.

Diversity is increasing by design as well. Organizations are recruiting a more diverse workforce to improve relationships with customers. A design team does not create a car for male buyers only; products must be sensitive to all potential customers. Increasing workforce diversity enables organizations to be more sensitive to the diverse markets in society. In addition, globalization is increasing diversity. As organizations become more global, the people in them must be able to interact in culturally diverse teams.

In addition to changes in the character of the workforce, organizations are making other changes that affect diversity. The greater use of teams increases the importance of diversity issues because representatives of the diverse work-force must interact with one another to perform their jobs. Teams must deal with the diversity that comes from differences in occupations, departments, and organizational statuses. The widespread restructuring of larger organizations has reduced the number of organizational levels. In these flatter organizations, more people interact as peers, as opposed to the more structured hierarchical relations. Work is more interdependent, making the ability to communicate with different types of people in new ways vital.

Types of Diversity

Although we often think of diversity in terms of gender or ethnicity, three types of diversity affect groups in organizations: demographic (e.g., gender, ethnicity, age), psychological (e.g., values, personality, knowledge), and organizational (e.g., tenure, occupation, status). Table 13.1 presents an overview of these types of diversity. Demographic diversity relates to the social categories people use to classify others (McGrath, Berdahl, & Arrow, 1995). In our society, distinctions of gender, race, ethnicity, nationality, age, and religion are considered important in many situations, varying in importance depending on the society and era. For example, religion is a more important demographic variable in the Middle East than in the United States. Differences among European immigrants were considered very important in the United States during the early 1900s, but these are not viewed as culturally important differences today.

TABLE 13.1
Types of Diversity

Demographic	Psychological	Organizational
Gender	Values, beliefs, & attitudes	Status
Race & ethnicity	Personality, cognitive, &	Occupation
Nationality	behavioral styles	Department/division
Age	Knowledge, skills, & abilities	Tenure
Religion		

SOURCE: Adapted from McGrath, J., Berdahl, J., & Arrow, H. (1995). Traits, expectations, culture, and clout: The dynamics of diversity in work groups. In S. Jackson & M. Ruderman (Eds.), *Diversity in work teams: Research paradigms for a changing workplace* (pp. 17–45). Washington, DC: American Psychological Association.

Psychological diversity relates to differences in people's cognitions and behavior. There are three types of psychological diversity. People vary in their values, beliefs, and attitudes; they may be conservative or liberal, religious or not religious, risk oriented or risk averse. People differ in personality and behavioral styles. As discussed in earlier chapters, people may be competitive or cooperative, assertive or aggressive. Finally, people differ in task-related knowledge, skills, and abilities. Team members may be technical experts, have artistic skills, and/or be good communicators.

Organizational diversity is caused by differences in people's relationship to an organization. Factors such as organizational rank, occupational specialty, department affiliation, and tenure are examples of organizational variables. These variables primarily affect an individual's status in the organization, which has important consequences for how people communicate in teams.

In organizational settings, the different types of diversity are not easily isolated from each other. Teams often exhibit all three types of diversity. There is also confusion in the ways people categorize types of diversity (Cox, 1995). On the one hand, it makes sense to view demographic factors (e.g., gender) as different from organizational factors (e.g., status). On the other hand, one of the main effects of demographic differences is that we give more power and status to certain group members than to others. Demographic and organizational factors interact through the flow of power and status.

The types of diversity vary in how easily they can be observed. Diversity based on age, sex, or race can be considered surface-level factors, as opposed to deep-level factors such as psychological variables (Harrison, Price, Gavin, & Florey, 2002). Surface-level factors affect people immediately. People who are similar in surface-level factors are more likely to be initially attracted to each other and form stronger social attachments. Deep-level diversity takes time to recognize. Consequently, the effects of these deep-level differences on teams take time to develop.

How Diversity Affects a Group

There are two ways to view how diversity affects a group (McGrath et al., 1995). The trait approach assumes that diversity affects how people act. In other words, people with different backgrounds have different values, skills, and personalities, and these differences affect how they interact in a group. The expectations approach focuses on the beliefs people have about what other people are like. These expectations or beliefs change how they interact with people from different backgrounds.

As people work together in a group, they develop a sense of identity with the group, which becomes stronger as the group becomes more cohesive and members establish working relationships with one another. Over time, members develop emotional bonds, create a common language for communicating, and share experiences. This leads to a convergence of attitudes, beliefs, and values that reduces the importance of background differences among team members (Harrison et al., 2002).

Although continued interaction affects some types of diversity, it does not affect all types. Interaction does not change people's personalities, their specialized skills, their races, or their ages. However, it does not have to change these characteristics to reduce the impact of diversity. People identify with a group to the extent that membership is emotionally important to them and they care about the collective goals of the group (Brewer, 1995). One implication of team formation is that team members shift their social categories and create a new social identity. Members working together in a team develop the category of teammate, and this category can become emotionally more important than the other ways of categorizing the people on the team.

2. Causes of Diversity Problems

There are several ways of viewing the causes of diversity problems. One view sees diversity as arising from our cognitive processes; it is an artifact of our need for social classification. This misperception creates interpersonal problems in the group. A special case of this is misperceptions by the team leader. An alternative view sees diversity as due to power conflicts arising from intergroup competition. Rather than being caused by psychological issues, diversity problems reflect competition and power struggles between groups.

Diversity as a Cognitive Process

Diversity is a social construction based on our cognitive processes. People categorize their social world into groups and treat the members of those groups differently based on their categories (Wilder, 1986). These categories are relatively arbitrary. For example, we are more likely to categorize people in ways that are easily observable (e.g., race rather than religion). Once these categories are formed, they have important implications for how people perceive and interact with others.

Social perception is the process of combining and interpreting information about others. The primary reason people categorize others is to simplify the world (Srull & Wyer, 1988). Dividing people into categories or groups makes it possible to predict what they are like. It is a simplification, often not very accurate, but an unavoidable component of human cognition.

The problem with social perception is that it leads to premature judgments about what others are really like.

Stereotypes are cognitive categorizations of groups that describe what people in the groups are like. Stereotypes may be positive, negative, or both. People may believe that engineers are very analytical, and this may be a good or bad attribute depending on the context. Stereotypes make people in a category seem similar to one another, yet different from us (Wilder, 1986).

This social perception and categorization process, by itself, is not bad. It helps people interact with others. The problem is that the process creates inaccuracies and biases that lead to misperceptions. Table 13.2 shows a set of common perceptual biases that negatively affect how we perceive others. From these biases, it is easy to see how our social perceptions can go wrong.

TABLE 13.2

Perceptual Biases—The Common Ways People Misperceive Others

Fundamental Attribution Error	The tendency to explain why someone is acting in a particular way by using personal rather than situational explanations. We tend to explain other people's behavior using personality traits and demographic variables, rather than looking at situational causes.
First Impression Error	The tendency to base judgments on our first impressions and ignore later information that contradicts them. Once we form a positive first impression, we create the circumstances to justify it.
Halo Effect	Once we have an overall positive or negative impression of someone, we assume that they are good (or bad) at everything. For example, if we like someone, we often assume they are competent and dependable.
Similar-To-Me Effect	The tendency to view people who are similar to you in a positive light.
Selective Perception	The tendency to focus on and remember only information that confirms our beliefs and ignore information that contradicts them.

SOURCE: Adapted from Greenberg, J., & Baron, R. (1997). *Behavior in organizations: Understanding the human side of work* (6th ed.). Upper Saddle River, NJ: Prentice Hall.

The problem of diversity is more than just categorization and perceptual biases. When people classify others, they divide their social worlds into in-groups and out-groups. This cognitive distinction has an emotional component (Tajfel, 1982). The group one belongs to is viewed more positively, and we like, trust, and act friendlier toward in-group members. In addition, similarity in demographics is often assumed to make for similarity in other areas, such as values, beliefs, and attitudes. Consequently, people are more likely to believe that in-group members are better than out-group members.

The addition of an emotional component to our categories shows how stereotypes become prejudices and discrimination. Prejudice is an unjustified negative attitude toward a group and its members. Prejudices typically are based on stereotypes. Prejudices may lead to discrimination if there is organizational and social support for the discrimination.

From this cognitive perspective, diversity is a cognitive categorization process that has an emotional component. The problem of diversity is that we misperceive people. People prejudge others on the basis of their categories, rather than how others actually behave. This causes people to treat others inappropriately, to have poorer communication due to their misconceptions, and to dislike and distrust others without getting to know them (Mannix & Klimoski, 2005).

Team Leader

One way diversity affects a team is through the relationship between a team member and a team leader (Tsui, Xin, & Egan, 1995). As was noted in Chapter 10, leader-member exchange theory describes the dynamics of this relationship. According to the model, the team leader decides early in the relationship whether the team member is part of the in-group or the out-group. If part of the in-group, the team member will receive more resources, mentoring, and assistance, have better performance evaluations, and be more satisfied with being part of the team than will members of the out-group.

There are two important issues to note in this description of the leader-member exchange. First, the in-group/out-group evaluation occurs very early in the relationship, before the leader actually knows much about the performance of the team member. Second, the impact of this early impression has long-lasting effects on the relationship.

A leader's quick decision that a team member is either in-group or out-group often stems from the perceptual biases listed in Table 13.2. During the initial interaction with the leader, the team member is categorized (first impression error). The leader is more likely to rate favorably a team member who is similar to the leader (similar-to-me effect). The leader is more likely to rate a team member favorably on many issues if the member's overall stereotype is positive (halo effect). Subsequent interactions rarely alter this first impression because information that supports the impression is remembered, while conflicting information is ignored (selective perception).

Diversity as a Social Process

An alternative to the cognitive view is that diversity problems arise from social competition and conflict. Why are gender and ethnicity important ways of classifying people? To a sociologist, it is because women and minorities are challenging the power position of white males in our organizations and society. Women and minorities are competing for scarce resources (e.g., jobs, office space, project resources) that the majority group wants to control.

When groups compete, their members form prejudices against each other. As noted in Chapter 5, when competing groups are united by common goals, these prejudices are reduced. A person can classify his or her social world in a variety of ways, but prejudices arise when the out-group is perceived as a threat to the individual's resources or power.

Diversity primarily affects power in a group (McGrath et al., 1995). Diversity affects group interaction primarily by creating power differentials within the group. Many of the negative effects of diversity may be viewed as the effect on the group of unequal power. As discussed in Chapter 8, unequal power in a group disrupts its communication process. In groups with unequal power, the level of communication is reduced and the powerful members control the communication process. Power differences affect group cohesion because individuals with similar status are more likely to interact with one another and form friendships (Tolbert, Andrews, & Simons, 1995).

3. Problem of Diversity

Analyses of diversity show no consistent impact, either positive or negative (Mannix & Klimoski, 2005). In diverse teams, members have different approaches to problems and access to different sources of information. This should help improve team performance, but only if the team uses these task-relevant differences. It is insufficient to have a diverse team; task-relevant differences among members must be used to gain the positive effects of diversity. Unfortunately, diversity may lead to misperceptions that reduce communication by minority members and increase emotional tension and conflict within the team. These effects prevent the team from fully using its resources.

Misperception

False stereotypes and prejudices of team members cause diversity problems. People from different backgrounds hold different values and respond to situations differently. These differences in values and behavior can be threatening to people's sense of what is appropriate. To deal with the psychological anxiety, people often either ignore or misinterpret the actions of people who are different.

These perceptual biases may cause people to discount the contributions of minority members. (Minority in this case means people with different demographic or organizational backgrounds from those of most group members.) Over time, minority members respond to this by contributing less to group communication. The lack of power these members experience causes them to have less impact on the group's decisions (Tolbert et al., 1995).

A potential benefit of diversity is to increase the types of information and variety of perspectives that may be used to analyze and solve problems in a group (Ven der Vegt & Bunderson, 2005). This benefit is lost if the group ignores the input of minority members or if the minority members do not provide input. The problem of diversity in teams occurs when the team overlooks the right answer because the "wrong" person came up with it.

Emotional Distrust

The dividing of a group into in-group and out-group members creates social friction. Power conflicts create a climate of distrust and defensive communication. Rather than forming a social unit, the group may become divided into cliques (Mannix & Klimoski, 2005).

These emotional issues create several group process problems. Diversity may lead to an increase in conflict because people are more distrustful. Not only are there more conflicts, but the conflicts are more difficult to resolve. Emotional distrust prevents the group from forming the social bonds necessary to form a cohesive group. Diversity may prevent the benefits of group cohesion from being realized.

Failure to Use Group Resources

Poor adaptation to diversity may cause a group to ignore and silence the views from the minority. This means the group is not fully using its resources. Diversity has other effects that limit the resources of the group. It may limit the careers of minority members. If minorities are not recognized and rewarded for their actual contributions, they may not become motivated team members.

The way the group treats minority members not only reduces their input in the group, but may reduce their desire to contribute. Over time, minority members become less committed to the team's goals and less motivated to perform for the team (Ancona & Caldwell, 1992). This in turn is used to justify not rewarding minorities or failing to provide them with opportunities and support to achieve more.

Diversity has been shown to affect turnover and socialization in work teams (McGrath et al., 1995). Minority members are more likely to have higher turnover in a team. It is easier for a team to socialize new members into the team if their characteristics are similar to those of the majority. However, a team that starts with a high level of diversity is more likely to have low minority turnover and have a less difficult time socializing diverse new members.

4. Effects of Diversity

The results of research on the effects of diversity on teams depend on how the research is conducted, the type of diversity examined, and the tasks the teams are performing. Sometimes organizations create diversity in teams on purpose to achieve a particular goal. Cross-functional teams are a type of diverse team used to deal with complex issues requiring a variety of skills or participation from various parts of an organization.

Research on the Effects of Diversity on Teams

Research on the effects of diversity on teams is contradictory. Part of the problem is the difference between short-term laboratory research on groups and the study of actual working teams. Homogeneous groups do better in the short run. However, many of the problems with diversity are related to miscommunication that goes away over time (Northcraft, Polzer, Neale, & Kramer, 1995). Another issue relates to the tasks the team is performing. Diversity is a benefit for some types of tasks, but a problem for others. Confusion is caused by the type of diversity being studied. Is diversity in demographic variables (e.g., gender, ethnicity) the same as diversity in personal variables (e.g., values, personality, skills)? Does it make sense to mix all types of diversity studies together?

Finally, it must be asked whether this is even the right question. Does it make sense to study the effects of diversity separate from an organizational context? The impact of diversity on a work team depends on the organizational climate and how the team manages diversity (Adler, 1986). Diversity is a fact of life for most organizations. The important question is not whether diverse groups are better or worse, but rather how to make diverse groups operate more effectively.

There is a growing body of research from work groups on the effects of diversity. This research typically compares homogeneous and heterogeneous (diverse) work groups. Jackson (1992) conducted a meta-analysis of this research that can be used to understand the impact of diversity on work groups. Her analysis categorizes the research along two dimensions: type of diversity and type of task.

One of the most basic divisions of diversity is by personal attributes and functional attributes (Table 13.3). Personal attributes include differences in personality, values, attitudes, and various demographic variables (e.g., age, gender, race). Functional attributes concern knowledge, abilities, and skills relating to the work environment.

The impact of types of diversity interacts with the type of task being performed and the group process. Tasks can be performance, intellective, or creative/judgmental. Performance tasks require perceptual and motor skills, which typically are evaluated using objective standards of quality and

TABLE 13.3

Impact of Diversity on Work Groups

	Type of Diversity	
	Personal Attributes	*Functional Attributes*
Type of Task		
Performance	Mixed effects	Diversity is beneficial
Intellective	Limited research evidence. Diversity may be a benefit	Limited research evidence
Creativity & Judgment	Diversity is beneficial	Diversity is beneficial
Group Process		
Cohesion & Conflict	Diversity is a problem	Limited research evidence

SOURCE: From Jackson, S., Team composition in organizational settings: Issues in managing an increasingly diverse workforce. In S. Worchel, W. Wood, & J. Simpson (Eds.), *Group process and productivity.* Copyright © 1992 by Sage Publications, Inc.

productivity. Intellective tasks are problem-solving tasks for which there are correct answers. These are the types of tasks that knowledge workers (e.g., engineers, teachers, accountants) perform. Creative/judgmental tasks typically are decision-making tasks that do not have correct answers. In addition to task performance, researchers examine the impact of diversity on the group process. Group cohesion and conflict are typical measures of the group process used in diversity research. The relationship between types of diversity and tasks is presented in Table 13.3.

When a group's outcome is studied, diversity has either a neutral or positive impact on performance for all types of tasks. It is only when the group process is used as the criterion that diversity has a negative impact. Although diverse groups may have more problems managing the group process, this does not seem to prevent them from performing better than homogeneous groups.

Positive benefits accrue when a team learns how to manage the challenges created by diversity (Mannix & Klimoski, 2005). Diversity can improve problem solving by increasing the number of perspectives on the problem. A diverse team is likely to have a greater variety of interpersonal relationships that provide avenues for more information gathering and assistance. Diversity in top management teams is related to innovativeness and willingness to make strategic changes in how the organization operates.

Diversity in personal and demographic attributes has mixed effects on the group, whereas diversity in functional attributes has primarily positive effects. Diversity in skills and knowledge helps a group to be better at problem solving (McGrath et al., 1995). Diversity in personal and demographic attributes can increase group conflict that disrupts the group's ability to reach consensus. For work teams, the most disruptive factor is often differences in organizational tenure or seniority (Mannix & Klimoski, 2005). In groups that have been working together longer, the link between diversity and conflict is reduced.

For example, Baugh and Graen (1997) examined the effects of gender and racial composition on perceptions of team performance in cross-functional teams. Members of heterogeneous teams rated their teams as less effective than homogeneous teams. However, outside evaluations of team performance found no difference between heterogeneous and homogeneous teams. This shows that heterogeneity may lead to less pleasant working relationships, but it does not affect team performance.

The clearest benefit of diversity is in creative and judgmental tasks. Although diversity may create conflict in judgmental tasks and make a group less efficient, diversity and conflict improve the quality of the judgments. Many different types of diversity can improve the creative performance of a team (Amabile, 1983). Diversity in cognition, age, organizational tenure, and education all have been shown to be relevant to creativity. The value of this diversity increases for "identity-relevant" tasks such as designing a marketing program that attempts to reach different groups in society.

Cross-Functional Teams

In most cases, diversity is something that happens to a group. The group members who come together to complete a task may or may not be a diverse set of individuals. However, there are cases in which diversity is created for a purpose. Cross-functional teams are a good example of diversity by design.

The complexity of modern organizations and the tasks they perform often require cross-functional teams (Northcraft et al., 1995). For example, when an organization designs a new product, the design team often includes members with different technical skills (e.g., electronics, materials, programming) because of the complexity of the product. The design team may include members from different departments (e.g., marketing, engineering, manufacturing) to ensure that the innovation is supported by the entire organization. The diversity in cross-functional teams is created by differences in knowledge and skills as well as differences in organizational position, occupation, and subculture. These two types of diversity (functional and organizational) create both benefits and problems.

Teamwork can be evaluated by examining effectiveness and efficiency. Effectiveness relates to a team's ability to complete its task; efficiency relates to performing the task with the least amount of time, labor, or resources. Diversity in a cross-functional team increases its effectiveness because the team is more creative and better able to implement decisions. Diversity has mixed effects on efficiency. It is a benefit to efficiency because people can be assigned to perform tasks that they are well qualified to perform. However, diversity may increase the amount of conflict in decision making and the amount of group time spent coordinating and performing tasks.

Although there is a need for organizations to use cross-functional teams, these types of teams can be difficult to establish and operate (Uhl-Bien & Graen, 1992). The diversity that makes a team so valuable to an organization makes it difficult for the team to operate successfully. The positive task-oriented conflicts that make the team valuable make communication and emotional conflict problems more likely to occur. A successful cross-functional team is like a successful negotiation; the participants retain their individual values and differences while forming an agreement that uses these differences in a synergistic way.

One of the challenges of cross-functional teams is managing conflict. Some conflicts arise from legitimate organizational or professional differences (Pelled, Eisenhardt, & Xin, 1999). Resolution of these conflicts is part of the value of cross-functional teams. However, other conflicts are related to stereotyping, distrust, and biases that limit communication among team members. Such biases prevent teams from negotiating agreements even when the agreements are in everyone's best interest.

5. Application: Managing Diversity

Organizations use a variety of approaches to manage diversity. Diversity problems are caused by misperceptions and prejudices, which create communication and group process problems. Diversity also is about competition and power conflicts. Both of these views of diversity are important and must be addressed. Programs to manage diversity focus on increasing awareness to eliminate misperceptions and biases, improving communication and group skills, and dealing with group and organizational issues.

Increasing Awareness

Organizations try to deal with diversity issues through training programs to increase awareness. Awareness programs are designed to increase understanding of diversity issues by making people more aware of their assumptions and biases about other groups. The goal is to increase knowledge and awareness of the issues, challenge existing assumptions about minority groups, and eliminate stereotypes (Battaglia, 1992).

In some cases, diversity training may backfire (Gardenswartz & Rowe, 1994). Diversity training can heighten emotional tension because it makes people feel uneasy; it can lead to a polarization of attitudes about minorities; and any blaming and personal attacks that occur during training may leave lingering emotional conflicts.

Awareness training needs to go beyond just teaching about cultural differences (Triandis, 1994). Increased understanding should lead to the development of social contacts and friendships that cut across demographic boundaries. It is these informal social contacts that develop into relationships that reduce misperceptions, lead to improved understanding of differences, and promote trust. To build this bridge among team members, the leader and the team should emphasize the similarities among members and develop a team culture that spans the differences (Mannix & Klimoski, 2005). Actions as simple as discussing what members have in common and their unique contributions are a way to start the bridging process.

Improving Skills

Many conflicts in diverse groups are due to miscommunication caused by stereotypes and distrust. To deal with them, team members can be trained to communicate better with one another and to appreciate the unique contributions of other members (Northcraft et al., 1995). Skill-based diversity programs improve people's interpersonal skills to better manage diversity issues. These skills include improved communication and facilitation skills that can be used to resolve misunderstandings (Battaglia, 1992).

A team leader may use a variety of techniques to improve diversity relations in the team (Armstrong & Cole, 1995). Developing agreement on the team's purpose, norms, roles, and procedures improves communication among members. The team leader should support communication from all members and obtain the commitment of all members to the team's goals. The team leader should clarify miscommunications during team interactions. If the team is having trouble with open discussions, the leader should use procedures to structure the communication. Finally, team leaders should encourage face-to-face interactions to develop social relations.

One of the disappointing findings about group decision making is that teams tend to focus on shared information during discussions, rather than the unique information held by individuals (Mannix & Klimoski, 2005). Although diverse teams have the potential to perform better than homogenous teams, this will only occur if the team can gain access to their unique contributions. The more successful the team is at creating an open communication climate that promotes trust and provides support to members, the more willing members are to risk stating their unique information on a topic.

It can be especially difficult for a "minority" team member to speak up in a team if that individual is alone (Mannix & Klimoski, 2005). The influence of minority opinions is increased with even limited support. This prevents the group from ignoring the information as idiosyncratic. The leader should make sure the team hears the minority view by creating appropriate communication norms and climate.

Improving Organizational Issues

One approach to managing diversity is to break down the social boundaries between people. This will not necessarily be accomplished just by having members interact in a team. There must be approaches that equalize power in the team for communication to break boundaries (Nkomo, 1995). One way to do this is to structure communication to equalize power and participation among members.

Developing superordinate goals or strong collective team identities can help diverse teams work more effectively together (Van der Vegt & Bunderson, 2005). When the team has a strong sense of team identity, members are more willing to share ideas and pay attention to the ideas of other team members. Team identification helps members move beyond individual differences and focus on the needs of the team. In addition, the team's work can be structured to require high levels of interdependence such that coordination and interaction are needed to complete the project (Harrison et al., 2002).

Diversity should not be linked to task assignments (McGrath et al., 1995). A team should not assign Asian members the technical issues, women the communication functions, and younger members the computer tasks. When tasks and stereotypes are linked, stereotypes and prejudices become a rational way of explaining what happens in the team.

In some cases, the problems created by diversity are related to the performance evaluation and reward system. When members of a team are not working cooperatively, the evaluation and reward system is one of the first places to look for a cause (Northcraft et al., 1995). Members of the leader's in-group are more likely to receive better performance appraisals and rewards. In a cross-functional team, members are often evaluated and rewarded by the departments they represent rather than by the team. Given this situation, their commitment to the team's goals is limited. Such conflicts are primarily about the organization's reward system, not about diversity.

Summary

Diversity is increasingly relevant for organizations because of the increased numbers of women and minorities in the workplace, the desire for workforces to reflect the diversity in society, and changes in how people work together. Although we often think of diversity in terms of demographic differences (e.g., gender, ethnicity, age), diversity also includes psychological (e.g., personality, values, skills) and organizational (e.g., department, status) differences among people. The impact on groups of diversity can be caused by differences among types of people or expectations about differences that cause people to treat others differently.

The problems created by diversity have several causes. People categorize others and use stereotypes to explain differences between groups. The categorization process can lead to misperceptions and cognitive biases. Team leaders are affected by these biases and may treat team members differently because of their backgrounds. Diversity may be due to competition and conflict between groups. The problems of diversity are related to attempts by majority groups to maintain their power and status.

The biases created by diversity may cause members of a group to misperceive and discount the contributions of minority members. This reduces minority members' ability to contribute to the group's efforts. Emotional distrust leads to defensive communication and power conflicts in the group. These factors disrupt the operation of the group and reduce the motivation of minority members to participate.

The effects of diversity on groups are complex. The performance differences between homogeneous and heterogeneous groups depend on the types of diversity and tasks. For personality and demographic diversity, the effects of diversity are mixed for tasks but are negative for group cohesion and conflict. For functional diversity (which relates to differences in skills), the effects of diversity are positive for most types of tasks. Cross-functional teams are an example of diversity purposefully created by organizations. These teams can be very effective but also may be difficult to establish and operate.

Organizations can develop programs to help groups better manage diversity issues. Diversity programs are designed to increase awareness of the differences among types of people, to improve groups' ability to communicate and resolve conflicts, and to create goals and organizational reward systems that encourage working together.

TEAM LEADER'S CHALLENGE—13

You are the professor in an undergraduate engineering design class. The year-long class uses student teams to complete a complex design project. Your goal is to simulate a real-world professional experience in the class, but you also need to ensure that it is a safe and productive learning experience for the students.

Like many engineering classes, there are few women students. In the past, you have not been concerned about gender issues when assigning students to project teams. However, last year you received several complaints from women students about feeling bullied and unsupported in their teams. These students were the only women in their design teams.

How should you distribute the few women engineering students among the project teams?

Are there actions you could take to provide support for the women students in the class?

How do you handle complaints from women students about team relations?

ACTIVITY: UNDERSTANDING GENDER AND
STATUS DIFFERENCES IN A TEAM

Objective. Diversity can be caused by demographic (e.g., gender) or organizational (e.g., status) differences. The more powerful group is more likely to communicate, speak more forcefully, and contradict others. It is sometimes assumed that women's communication will be more polite and deferential than men's communication, but this may have more to do with status than gender. This activity will help to explore this question.

Activity. Use the observation form (Activity Worksheet 13.1) to record the communication in a team meeting that contains male and female members. Or organize a small group discussion on the Team Leader Challenge with a mixed gender group.

ACTIVITY WORKSHEET 13.1
Observing Team Communication Differences

	Team Members					
	1	2	3	4	5	6
Phrases ideas tentatively and politely						
Shows agreement and support for others						
Confronts issues using direct and forceful language						
Contradicts and disagrees with others						
Total number of communications						

Analysis. Women and low-status team members use the first two communication styles more often, while men and high-status team members use the last two communication styles more often. Compare the various communications of women with that of men in the group, and compare the communication level and style of high-status (e.g., leader) with that of low-status members. Also note which type does most of the communicating in the group.

Discussion. How do you explain the differences among communication styles of team members? Are these differences due to status, personality, or gender differences? What should a team do to make sure diversity differences do not interfere with full participation and acceptance in the team?

PART IV

Teams at Work

14

Team and Organizational Culture

The shared values, beliefs, and norms of a team or organization are known as its culture. A team's culture affects how team members communicate and coordinate work. Organizational cultures vary in the ways individuality, status, and uncertainty are accepted and used. These organizational cultural differences affect how teams operate within organizations.

The success of work teams depends on the level of support they receive from their organization, which in turn depends on the organization's culture. At the same time, the use of teams changes an organization's culture. Teams need to be sensitive so that they do not violate the existing organization's culture, but they also should encourage changes to support teamwork.

Transnational teams are composed of members from different cultures who are linked by communication technology. These teams must manage cultural differences to operate effectively.

Learning Objectives

1. What is a team's culture and how is it created?

2. How does a team's culture affect its performance?

3. Understand the concept of organizational culture and the effects it has on people in organizations.

4. How do the different dimensions of organizational culture affect teamwork?

5. How are U.S. and Japanese teams different?

6. What are the two main types of organizational cultures that affect the use of teams?

7. How can organizational cultures be changed to support teamwork?

8. What strategies can transnational teams use to manage cultural differences?

1. Team Culture

A team's culture is the shared perception of how the team should operate to accomplish its goals. Team norms, member roles, and patterns of interaction are included in the team culture. Teams do not develop their culture from scratch; they incorporate cultural norms and values from their organization and society (Wheelan, 2005). Agreement about norms, values, and roles (about the team's culture) reduces anxiety and improves communication in a team.

A team's culture and its norms often develop through precedent (Thompson, 2004). Behavior patterns that emerge early in a team's life define how the team will operate in the future. These norms affect the team's performance. The leader also plays a key role in the development of a team's culture (Schein, 1992). Leaders should try to establish an appropriate culture early in the team's life, because it is easier to start a culture than it is to change an existing one.

The main cultural values of work teams are commitment, accountability, and trust (Aranda, Aranda, & Conlon, 1998). Commitment relates to both the task and the other members on the team. Commitment indicates the willingness to participate and become involved in the task and to support other team members. Accountability delegates responsibility for the team's success; this may reside with the individual, the team, or the organization's hierarchy. A requirement for accountability is empowerment. The team cannot be held accountable if it does not have the authority and power to act on its own. The third aspect is trust. Without a culture of trust, the team cannot have free and open communication.

Team culture relates to members' willingness to share information, the level of mutual support, and how the team uses training. The culture of a work team is related to the culture of its surrounding organization.

A benefit of teamwork is the sharing of information among team members to coordinate work activities (Zarraga & Bonache, 2005). This knowledge-sharing requires a collaborative team culture. Team members may be reluctant to share information given that information is a valuable individual commodity in work organizations. A collaborative team culture includes mutual trust, active empathy, leniency in judging others, courage to state opinions, and willingness to help. When a collaborative culture exists, the team is better able to use the resources of its members.

Team culture relates to team support or the availability of helping behavior within a team (Drach-Zahavy, 2004). Support includes both emotional support and help or assistance in performing the task. The level of team support relates to the leader's behavior and the team culture. Role modeling by the leader and other team members encourages team members to provide one another with support. Teams that are group-oriented (rather than individualistic) and are less status-oriented provide more team support. Studies on high-stress action teams such as nursing teams show that supportive team cultures reduce the negative impacts of work stress (Drach-Zahavy, 2004).

Team culture has an impact on whether training is used by a team (Smith-Jentsch, Salas, & Brannick, 2001). Team culture relates to the degree to which the team expects and supports attempts to try newly learned skills. Culture does not affect what is learned in training, but whether the newly learned skills are applied. For example, acting assertively is highly dependent on the team culture established by the leader. If the team leader rewards assertiveness, team members are more likely to act assertively after receiving communications training.

For many work teams, the team's culture is a reflection of the organization's culture (Thompson, 2004). Companies that are successful using teams often have norms that foster participation and innovation. However, culture is more a property of the team than it is of the entire organization. Teams may have cultures and norms different from those of the larger organization. For example, the Hawthorne studies showed that work groups in the same area of the factory could have different team cultures leading to different performance levels (Sundstrom et al., 2000).

2. Defining Organizational Culture

The concept of organizational culture arose during the 1980s, partly as a result of the many comparisons between U.S. and Japanese organizations. Peters and Waterman (1982) used the concept of organizational culture as a way of describing the practices of the best U.S. companies. Schein (1992) was influential in arguing that the principles for examining national cultures could be used to describe organizational cultures.

Organizational culture refers to the shared values, beliefs, and norms of an organization. Researchers studying organizational culture emphasize various aspects as key to understanding how organizations operate. For Deal and Kennedy (1982), an organization's customs, rituals, and traditions help reveal the underlying values that guide organizational decision making. Davis (1984) focuses on the shared meanings and beliefs of organizations because they affect an organization's strategies and operating procedures. Kilmann and Saxton (1983) view culture as determining the group norms and behavioral patterns of employees.

Regardless of which characteristic of organizational culture is selected, there are features of cultures that are common to all perspectives (Schein, 1992). All members of an organization share its organizational culture. Culture provides structural stability for the organization because its influence is pervasive and slow to change. The varied aspects of culture are integrated and form a consistent pattern. Culture reflects the shared learning by organization members that contains cognitive, behavioral, and emotional elements. Finally, organizational culture affects both the internal operations of the organization and how it relates to its external environment.

Teams and organizational culture have a mutually influencing relationship. Teamwork occurs more easily in some types of organizational cultures than in others. For example, cultural norms about power and control affect the way communication flows among levels in an organization (Zuboff, 1988). New team-based work practices such as total quality management may be unsuccessful because they contradict cultural norms about communication and power (Bushe, 1988). The use of teams changes the ways people work and relate to one another, which changes the organizational culture. Over time, an organization's work systems tend to become congruent with its organizational culture.

Organizations are not uniform, and they do not necessarily have uniform cultures. Rather than viewing organizations as having single cultures, organizations may be viewed as containing networks of groups that develop their own styles of operating and interacting. These are organizational subcultures. Such subcultures arise from mergers and acquisitions, geographic differences in facility locations, or occupational areas in an organization.

Organizations may be characterized by how integrated their separate subcultures are. In strong organizational cultures, the networks are strong and the organizations have dominant cultures; in weak organizational cultures, the subcultures are relatively independent (Van Maanen & Barley, 1985). When the shared beliefs and assumptions held by working groups are similar across organizations, the organizations have strong cultures. For example, Hewlett-Packard has a strong organizational culture that defines how managers should treat employees: open-door policy, fostering independence, and promoting equal-status relations. These practices operate throughout all divisions of the company.

One important cause of subcultures is related to an employee's occupational community. An occupational community refers to the shared knowledge, language, and identity formed by working in a particular area of specialization (Schein, 1992). For example, engineers and sales representatives have different occupational communities and therefore occupy different subcultures within an organization. Even in strong organizational cultures such as Hewlett-Packard, people from engineering and sales use different professional languages and have different styles of interacting with others. This can make working together on cross-functional teams difficult (Adler, 1991).

3. Dimensions of Organizational Culture

A number of approaches are used to determine the dimensions of international and organizational cultures. Dimensions allow the establishment of frameworks that can be used to compare different cultures. Triandis (1994) reviewed several approaches used by anthropologists. Hofstede (1980) studied the employees in international companies to compare national differences. Aranda and colleagues (1998) examined work teams in different organizations in the United States. Each of these approaches has identified a similar set of key dimensions.

Organizational cultures can be compared on three dimensions: individualism versus collectivism, power, and uncertainty (Table 14.1). The following subsections examine these three dimensions and show how they can be used to compare teams in the United States and Japan.

Individualism Versus Collectivism

In an individualist culture, people have loose ties with one another and expect to be responsible for themselves and their immediate families. People

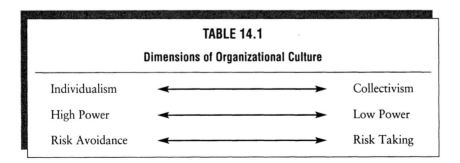

TABLE 14.1

Dimensions of Organizational Culture

Individualism	← — — — →	Collectivism
High Power	← — — — →	Low Power
Risk Avoidance	← — — — →	Risk Taking

seek individual achievement and recognition and might have trouble committing to team goals. Collectivist cultures value the ties between people. People are expected to look after one another. Self-interest is subordinate to the interests of the social group or team.

This distinction is reflected in how people respond to conflict and conformity pressure. Collectivists value compliance because the harmony of the team is considered vital. Conformity is expected; open conflict is discouraged. Individualists often enjoy conflict, competition, and open controversy. They seek argument for its own sake. A successful team finds ways to encourage participation of compliance-oriented members, who may be shut out of the team's decision-making process by its conflict-oriented members.

This dimension has mixed effects on teams. Collectivists are easier to organize in teams because they value cooperation, but a team may run into problems because conformity pressure prevents it from achieving the open communication and constructive conflict that is one of the benefits of teamwork.

Although collectivists value cooperation, this does not mean they cannot be competitive when it is appropriate. Japanese business organizations are collectivist. Cooperation among employees is highly valued and rewarded. However, they are very competitive toward outside organizations.

Power and Status

The power dimension is the degree to which unequal power is accepted by people in a culture. In high-power cultures, large power and status differences are acceptable. In such cultures, great respect and deference are shown to higher status people, and challenging authority feels uncomfortable. On teams, members are more willing to accept the leader's decisions.

In low-power cultures, people are less willing to accept the authority of others on the basis of the positions they hold in an organization. Their viewpoint is more egalitarian. Team members take initiative and do not automatically accept management directives.

A high-power culture can be a problem for a team. It makes team members more willing to accept the team leader's view, a lack of independence that reduces team creativity. Participation is not easy for people in high-power cultures because they believe communication from above is more important than their own ideas. In an unequal-status situation, the higher-status person does most of the communicating, and most communication is directed at the person with highest status.

A team in a low-power culture can be difficult to manage. Members' more open communication style creates more conflicts. Their sense of independence from the organization's authority may lead the team into decisions that are not sensitive to the politics of the organization. Egalitarian communication in decision making may improve the quality of the decisions but reduce the team's ability to implement them because of lack of deference to the concerns of the surrounding organization.

Uncertainty and Risk Avoidance

Cultures vary in how much people are willing to accept uncertainty and in their desire to avoid taking risks. Uncertainty is the degree to which people feel threatened by and avoid ambiguous situations or change. In risk-avoidance cultures, social harmony and stability are valued. People want to have rules and norms that define appropriate behavior, and they prefer things to stay the same so they know what is expected of them.

Risk-avoidance cultures value social harmony more than change. Open conflict is considered inappropriate; people avoid controversies or become compliant during controversies. People in risk-avoidance cultures try to maintain the security of the status quo, in part because they fear the potential for failure during change.

Risk-taking cultures value change. They tend to be action-oriented rather than planning changes in advance. People in these cultures are open and willing to try out new ideas. Conflict is more likely to be viewed as positive because it encourages new ideas and change.

Comparing the United States and Japan

The United States and Japan make an interesting comparison of the effects of culture. Applying the dimensions discussed above, U.S. organizations tend to display individualism, low power, and risk-taking, whereas Japanese organizations tend to display collectivism, high power, and risk avoidance. These differences significantly affect the use of teams and how teams operate.

The focus of U.S. management practice is on controlling, motivating, and rewarding individual performance. The individual remains independent of the organization and is expected to remain committed to the organization only so long as it is in his or her best interest. There is less use of teamwork in the United States than in other industrialized countries (Cole, 1989). The focus on competition and individualism in U.S. culture limits teamwork, especially among professional and managerial staff.

In the Japanese approach to management, the individual does not have a job, but rather has become part of the organization (Ouchi, 1981). Japanese organizations stress the interdependence of all employees. Their participative style is marked by mutual respect and common interests (Pascale & Athos, 1981). Consensus decision making is practiced at all levels of organizations.

Teamwork programs such as quality circles and production teams are more common in Japanese companies. Management creates these teams to serve as mechanisms for employee participation. The focus of teamwork is on improving the productivity of the work system. Participation allows employees to make suggestions, but management retains control over all decisions (Cole, 1989). Unlike teamwork programs in the United States, participation does not imply power-sharing with the workers in hierarchical Japanese corporations.

The Japanese have a more cautious view of change because their culture is more attuned to the value of promoting social harmony. Japanese companies tend to implement incremental changes because of their concern for social relations and job security (Prochaska, 1980). On teams, there is less conflict and more conformity. Consequently, Japanese teams are viewed as less creative and less willing to take risks than U.S. teams.

Japanese teams use consensus decision making. Consensus is easier to reach in Japanese teams because people are less independent and try to avoid conflict. They are confident that compromise solutions will be found, so they do not rush to decisions. This makes Japanese decision making slower, but the implementation of decisions is faster. Once a Japanese team has made a consensus decision, it knows that everyone will support the implementation of the decision.

This comparison between U.S. and Japanese cultures demonstrates several points about organizational culture. First, cultures affect how teams operate. Second, national cultures affect organizational cultures. Third, cultures do not prevent the use of teams, but they do create an environment for teams that affects the way teams operate.

4. Organizational Culture and Teamwork

An organizational culture that encourages employee involvement and participation is a necessary support for teamwork and self-managing teams. In a supportive organizational culture, managers are less likely to resist using teams, and there are better relations between teams and other parts of the organization. Self-managing teams are much more likely to be successful in organizations in which the organizational culture supports empowerment and teamwork. Overall, organizational culture is one of the largest predictors of the successful use of teams by companies (Levi & Slem, 1995).

Organizational culture defines the norms that regulate acceptable behaviors in an organization. When these cultural norms conflict with the use of teams, organizations have a difficult time using teams successfully. For example, norms about communication that are part of an organization's culture may limit the organization's ability to use teams. Many organizations do not have open communication from workers to managers, across departments, or from top management to the rest of the organization. This limits the types of communication that occur in a team and limits the team's ability to relate to other parts of the organization.

Thomsett (1980) identifies several features of a supportive culture for teamwork. First, the organization must believe that people are an asset to be developed. People should be grouped together so that they can use their skills and expertise to perform a task, rather than having their jobs broken down in such a way that only individuals can perform them. The organization should have a participative management structure with few organizational levels. The control system should emphasize self-regulation and be focused on commitment rather than on control. Finally, the environment should be non-hostile. Fear and anxiety, which result in employees being defensive and distrustful, drive hostile organizational cultures.

Walton and Hackman (1986) identify two distinct types of organizational cultures that affect the use of teams: control cultures and commitment cultures. (This distinction is similar to McGregor's [1960] Theories X and Y management.) Status and power drive the control strategy; it is hierarchical and tightly controlling. The relations among people are adversarial and untrusting. Teams are difficult to form and operate in this context. The commitment strategy reduces the number of organizational levels of authority, focuses on quality, and adopts methods to encourage open communication and participation. It uses teams to operate and gives them the autonomy and authority to operate successfully. This type of culture empowers both individuals and teams so as to increase commitment to their organizations' goals.

Obviously, most organizations fall somewhere in between these two approaches. Although managers may want to shift to a commitment strategy, if the existing culture is control oriented, it will be difficult to change. Developing teams is a struggle when their use is not compatible with the existing cultural practices. Teams operate better in a commitment-oriented culture because they are given the resources, training, and power they need to operate.

It is critical for the organizational culture to support collaborative work for teams to operate successfully (Dyer, 1995). It is an arduous task for a traditional control-oriented culture to change enough to promote good teamwork. Employees are often cynical of announcements by management that it will simply create teams when the culture does not support the use of teams.

Organizations want to change; they want to promote teamwork and create commitment-oriented organizational cultures. But they do not want to change the existing system of power, authority, and rewards. When these organizational system processes diverge from announced cultural changes, employees are rightly skeptical. Incongruence between the culture and the work system creates confusion.

Teams and Cultural Change

The importance of organizational culture as a primary support for teamwork is both a problem and a benefit (Levi & Slem, 1995). Organizational culture is not easy to change. Developing an organizational culture that supports teamwork is a long-term process; it is not something that can be dictated by top management or announced as a new organizational program. Changing an organizational culture requires a consistent effort on the part of management to show that employee involvement and teamwork will be valued and rewarded. This must be done through both communication and action. If what an organization says it believes in does not match its actual behavior, a credibility gap is created, and trust between organization and members drops.

In organizations without uniform cultures, cultural change may occur within subcultures (Dyer, 1995). However, even successful subcultures will not necessarily spread to other parts of the organizations. For example, General Motors created a new organizational culture at its Saturn facility to support teamwork. Although this team-based approach to manufacturing has been successful, the approach has not been easy for other GM facilities to adopt.

There is a benefit to this relationship between organizational culture and teamwork. Once an organization begins to create an organizational culture that supports teamwork, the culture will support a wide variety of teams and, eventually, the transition to self-managing teams. The organizational culture provides the foundation, and from that foundation it is able to experiment with developing the types of teams that can successfully fulfill its mission.

Implementing Teams in Difficult Organizational Cultures

Changing organizational culture is a difficult task in traditional work environments. It is relatively easy to identify those aspects of organizational culture that limit the use of teamwork. Once the factors are identified, many approaches may be used to change them. The problem is how to implement the approaches if they are incompatible with an organization's culture. One strategy for dealing with this problem is to examine the culture and use it to select organizational change strategies (Davis, 1984).

The relationship between organizational culture and the potential to adopt organizational changes is shown in Figure 14.1. To have the most impact, actions should be selected that are both important (will have a large impact) and relatively acceptable to the existing culture. When these actions are implemented, they cause the organizational culture to change. Eventually, other important changes become more acceptable to the new culture.

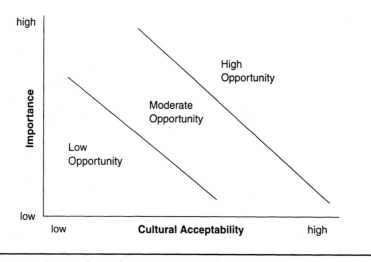

Figure 14.1   Potential to Adopt Organizational Changes

In some companies, a wide variety of organizational changes supporting teamwork is culturally acceptable (Levi, 1988). This allows change agents to make substantial changes to support the use of teams, such as changing work processes and reward systems. However, in other companies, most of the important changes are unacceptable to the organizational culture. If managers are reluctant to relinquish power and authority to teams, the changes needed to support teams cannot be implemented.

This cultural approach can be used to analyze the types of team decision making used by organizations. Organizations rarely shift from traditional management to self-managing teams because the amount of power-sharing in self-managing teams is not culturally acceptable. Instead, traditional organizations start by using teams with supervisors who use consultative decision making. Once the organizational culture accepts this approach, democratic decision making becomes more culturally acceptable. Eventually the use of self-managing teams may be acceptable to the organizational culture.

This incremental approach to changing organizational culture describes the evolution of teamwork in the United States. Teamwork programs during the early 1980s, such as quality circles, were experiments in using teams at work (Cole, 1989). As the use of these temporary teams became more acceptable, companies began to expand their use of teams by creating permanent work teams. Rather than just providing advice about quality issues, these permanent work teams were used to perform the day-to-day operations of companies. When companies learned to accept the new way of work, they began empowering the teams and shifting to self-managing teams.

5. Transnational Teams

Transnational teams are composed of individuals from different cultures working on activities that span national borders (Snell, Snow, Davison, & Hambrick, 1998). They are formed in a global company or through alliances among companies in different geographic areas. Transnational teams use representatives from multiple countries to ensure that the perspectives of local organizations, cultures, and markets are represented in the team. The main challenge for such teams is to learn how to integrate this cultural diversity into a functioning unit.

Transnational teams deal with three important concerns for global companies: local responsiveness, global efficiency, and organizational learning (Snow, Snell, Davison, & Hambrick, 1996). Their work serves to customize products and services to different cultures and coordinate local activities and markets for global companies. At the same time, transnational teams help integrate operations across parts of an international organization to improve efficiency. Transnational teams encourage innovation by bringing together ideas from various parts of an organization.

To be successful, transnational teams must deal with differences in culture that affect how people work and communicate in teams (Earley & Gibson, 2002). The two main cultural dimensions that affect these teams are individualism/collectivism and status, or power distance. They are more difficult to resolve because these teams rely on technology for their communications.

Many of the difficulties transnational teams experience stem from communication problems (Earley & Gibson, 2002). Cultures vary in how they communicate information. For example, in collectivist cultures, such as Asian countries, communication is often indirect with a positive tone. Communication frequently uses qualifiers and ambiguous words to avoid confrontation and preserve group harmony. In individualist cultures, communication is more direct, even when communicating negative information. Communication is about facts and is viewed as distinct from the relationship with the listeners.

Rewarding transnational team members may pose problems because of cultural ideas about how rewards should be given (Snell et al., 1998).

A focus on individual rewards may be considered inappropriate for team-work in collectivist cultures. Although team rewards are valuable, team members are often more responsive to the rewards they receive from their home organization.

Cultures vary in how status oriented they are (Earley & Gibson, 2002). In high status-oriented cultures, communication is more formal, with most coming from higher-status members, while lower-status members are polite and deferential in their communication. In less status-oriented cultures, communication is more information-oriented and participation is more equal.

Miscommunication in transnational teams is caused by cultural differences in the way people communicate; these differences are made worse by their reliance on communications technology, which makes it more difficult for transnational teams to develop trust and mutual understanding in their communications (Earley & Gibson, 2002). Culture-based communication problems become more difficult to manage in these virtual teams.

Strategies to develop effective transnational teams include spending more time initially to start the team, team training, and using strong leadership. The goal of these actions is to develop a hybrid team culture that will unify the team.

Transnational teams should schedule face-to-face meetings early in their existence to develop personal relationships and a shared understanding among team members (Earley & Gibson, 2002). Clearly shared goals, norms, member roles, and agreement about performance criteria should be established (Snow et al., 1996). These are part of a formal team contract that should be developed at the onset of teamwork to reduce misunderstandings later. Transnational teams should spend more effort developing project plans and other types of project management structures. Face-to-face meetings should be scheduled at key points in the plan to clarify any misunderstandings about the team's progress.

Training programs that explain the organization's strategy and culture encourage a common perspective (Snell et al., 1998). This type of training is especially useful at the onset of teamwork. Cross-cultural team-building that increases awareness of cultural differences in work practices and communication will improve team operations. Teamwork skills that should be the focus of training include conflict resolution, negotiation skills, project management, and interpersonal communication. In addition, training in the use of communications technology should decrease misunderstandings caused by differences in the use of technology.

Strong leaders are valuable for coordinating actions and managing conflicts in the team (Katzenbach & Smith, 2001). Strong team leaders are more acceptable in some cultures, such as Japan, so it may be hard to avoid using them. Virtual teams often have more powerful leaders to help coordinate communications and work assignments. Multicultural teams perform better with strong designated leaders (Earley & Gibson, 2002). Leaders provide direction, motivate team members, and ensure the team stays on course. Leaders help develop the hybrid culture that unites a diverse team.

Effective transnational teams often develop a strong hybrid culture that provides a common sense of identity for team members and facilitates their interactions. A hybrid culture is both a set of rules about how to act and a set of expectations about how the team operates (Earley & Mosakowski, 2000). It creates a shared understanding that allows members to better interpret communication from other team members. Cultural diversity is a benefit to teams because of the variety of perspectives and skills their members contribute, but teams need to develop their own hybrid culture to encourage collective effort. Because developing a hybrid culture takes time, team performance typically improves over time in transnational teams.

Summary

A team's culture is defined by its norms, roles, and values. Team culture develops over time but is strongly influenced by its organizational context. Culture has many influences on how a team operates because it affects commitment to the team, styles of communication and collaboration, and the support members provide for one another.

Organizational culture relates to the shared values, beliefs, and norms of an organization. It provides a sense of identity to its members and defines acceptable behaviors. Cultures vary in depth from surface rituals to deepseated underlying values. Organizations may have unified cultures or may be composed of networks of subcultures based on occupation or background.

Organizational culture may be viewed as varying along three dimensions: individualism, power, and uncertainty. The individualist-collectivist dimension defines how group-oriented and cooperative people are. The power dimension examines whether people accept power differences or strive for egalitarian relations. The risk dimension concerns whether people value rules and stability or are willing to take risks to change how they operate. U.S. and Japanese companies differ on these three dimensions of teamwork.

The use of teams in an organization depends on its organizational culture. Cultural norms can either support teamwork or limit a team's ability to operate effectively. Two distinct types of organizational cultures are (a) those based on power and control and (b) those based on participation and commitment. The two types provide very distinct contexts for teamwork.

The importance of organizational culture for teamwork is both a benefit and a problem. Once an organizational culture supports teamwork, it is usually able to support a wide variety of types of teams. However, organizational cultures that do not support teamwork are difficult to change. Although they can be encouraged to change incrementally to support teamwork, there is no guarantee that this approach will lead to environments that support the use of teams.

Transnational teams are composed of members from different cultures who deal with problems of global connectivity and local responsiveness for international companies. These teams must deal with cultural differences that affect communication and power dynamics while relying on technology for communication. Successful transnational teams tend to spend more time initially developing social relations and team practices to create a unifying hybrid team culture.

TEAM LEADER'S CHALLENGE—14

You are the manager of the sales staff at a consumer products store that is part of a national chain. The company is very hierarchical and operates following strict, bureaucratic procedures. It is a classic "command and control" organizational culture.

You are concerned that the strict focus on rules and procedures is hurting customer service and relations. Employees seem more concerned about following the rules than pleasing the customers. You believe that shifting to teamwork, with you as team leader rather than manager, would encourage more customer service orientation among the staff. However, you are uncertain whether teamwork is compatible with the organization's culture.

How can you create teamwork in an organizational culture that is not team-oriented?

What kinds of problems will you encounter using teams in this organizational environment?

How do you handle relations between the team and the larger organization?

ACTIVITY: EVALUATING A TEAM'S CULTURE

Objective. Teams have cultures that are similar to and different from the cultures of their organizations. It is essential to know both types of cultures because teams encounter difficulties when their cultures are too much at variance with the organization's culture. Cultures vary on the following four dimensions. In individualist cultures, people seek individual achievement and recognition, whereas collectivist cultures value the ties between people, and self-interest is subordinate to that of the group. In high-power cultures, people show great respect to higher-status people and feel uncomfortable challenging authority; low-power cultures take a more egalitarian view and are less willing to accept authority. In risk-avoidance cultures, people value stability and group harmony, whereas people in risk-taking cultures are action-oriented and willing to take risks. In control-oriented cultures, leaders attempt to monitor and control the behavior of subordinates, whereas leaders in commitment-oriented cultures are facilitators who guide and motivate subordinates.

Activity. Discuss with members of an existing team their team's culture and the culture of the organization of which the team is part. The existing teams can be either work teams or student teams at a university. Use the rating form (Activity Worksheet 14.1) to note the team and organization's culture on the four dimensions.

ACTIVITY WORKSHEET 14.1

Evaluating a Team and Organization's Culture

Rate the Team by placing a "T" on the scale below; rate the Organization by placing an "O" on the scale.

Individualism -- Collectivism

High Power -- Low Power

Risk Avoidance -- Risk Taking

Control Oriented -- Commitment Oriented

Analysis. How similar are the team's and the organization's cultures? On which dimensions do the team and organization have culture gaps? What problems can occur when there are differences between the team's and the organization's cultures?

Discussion. Why is it important for a team's culture to be compatible with its organization's culture? If the organization's culture has difficulty supporting teamwork, what should a team do?

15

Virtual Teams

Most of our experience with teams is through face-to-face interactions. However, virtual teams interact through communication technologies. There is a variety of technologies available for virtual teams. The use of technologies by teams changes how people interact and the dynamics of the group process. Virtual teams offer both benefits and challenges to organizations that use them. The available technologies are still evolving, and teams are still learning how to select, use, and adapt them to meet their needs.

Learning Objectives

1. What is the difference between the direct effects and secondary effects of a technology?

2. How do communication technologies support teamwork?

3. Understand the main characteristics used to analyze communication technologies.

4. How do communication technologies affect status, deindividuation, conformity, and miscommunication?

5. How can communication norms help to deal with the communication problems in virtual teams?

6. How are task performance and decision making affected by using communication technologies?

7. What is the impact of virtual teams on leadership in organizations?

8. What factors should be considered when selecting the right communication technology?

1. Use of Communication Technologies

During the past two decades, organizations have adopted a variety of new communication technologies. These new forms of organizational communication include electronic mail (e-mail), voice mail, videoconferencing, electronic bulletin boards, and intranets. The characteristics of the new communication media are similar to existing forms of organizational communication, but add new dimensions.

The use of communication technology has led to the creation of virtual teams. A virtual team is any team whose member interactions are mediated by time, distance, and technology (Driskell, Radtke & Salas, 2003). The core feature is not the technology, but that the team works together on a task while physically separated. Most large companies use virtual teams to some degree (Hertel, Geister, & Konradt, 2005). Virtuality is a matter of degree; most virtual teams have some face-to-face contact.

Communication technologies have had widespread effects on organizations (Axley, 1996). Technology directly affects work design, organizational design, and communication patterns, as well as secondary social effects caused by the reduced social and organizational cues in the messages (Sproull & Kiesler, 1991). The primary goals of virtual teams are to improve task performance, overcome the constraints of time and space on collaboration, and increase the range and speed of access to information (McGrath & Hollingshead, 1994). These goals are related to the direct effects of technology.

As with other types of technology, the largest effects are often the secondary effects, including unanticipated social and organizational effects. Because communication plays an important role in maintaining social relations and organizational culture, the lack of social information in a type of communication may prevent new social relations from developing and may not build a sense of community in the same way as traditional communication (Taha & Caldwell, 1993). Although communication technologies structure how people and groups communicate, people do not passively accept the technology's constraints. Instead, they adapt and modify the technologies to suit their needs (McGrath & Hollingshead, 1994).

The use of communication technologies has both positive and negative effects (Harris, 1993). Positive effects include improved speed and dispersal of communication, increased access to information, increased amount of communication, easier connections to others, and improved planning and decision making. Negative effects include information overload, inconsistent access, decreased face-to-face communication, disrupted organizational relations, and increased isolation.

Communication Technologies and Teams

Communication and information technologies can be used to support teamwork in four different ways (McGrath & Hollingshead, 1994; Mittleman & Briggs, 1999). First, technologies can gather and present information, such as collaborative document management systems and electronic whiteboards. Second, technologies help team members communicate both internally and with outside organizations. Third, information technologies help teams process information by providing systems to structure brainstorming, problem-solving, and decision-making activities. Fourth, technologies may be used to structure the group process through meeting agendas, assignment charts, and project management tools.

Communication technologies free teams from the constraints of time and place (Mittleman & Briggs, 1999). Their use creates new ways for teams to meet and interact. The options created by the use of information technology are presented in Table 15.1.

Same time, same place (STSP) meetings are the traditional face-to-face team meetings. Even when a team primarily interacts via technology, there is a value to face-to-face meetings. These are especially important when the team is formed to help establish social relations. Face-to-face interaction is important for celebrating major team milestones and dealing with major changes in the team's focus or task.

TABLE 15.1

Types of Meetings Created by Information Technology

Type of Meeting	Example
STSP—Same-time, same place meetings	Face-to-face meeting
STDP—Same-time, different-place meetings	Videoconferencing
DTSP—Different-time, same-place meetings	Computer databases
DTDP—Different-time, different-place meetings	Intranet bulletin board

SOURCE: Adapted from Mittleman, D., & Briggs, R. (1999). Communication technologies for traditional and virtual teams. In E. Sundstrom (Ed.), *Supporting work team effectiveness* (pp. 246–270). San Francisco: Jossey-Bass.

Technology does have an impact on team meetings. Technology-enhanced meeting rooms provide computers for all participants so they can interact directly with presentation materials. Computer decision support systems help structure meetings, especially for activities such as brainstorming and voting.

Same time, different place (STDP) meetings are distributed meetings in which team members interact through a combination of telephone, video, or text. Although videoconferencing is the most popular image of STDP technology, it is not necessarily the most useful technology. Video images of participants may be less useful than audio with shared data. Participants often prefer being able to manipulate data and images related to the task to focusing on images of the other participants.

Different time, same place (DTSP) meetings are useful for work teams that exist across different shifts, or for teams whose members travel frequently or telecommute. The information technology serves as a storage system so members can pass on information as needed. Project management and other software systems can create a framework for noting the status of a project or an agenda.

Different time, different place (DTDP) meetings are those in which team members share the same virtual space on an intranet (i.e., internal communication network). Technologies such as online bulletin boards, chat rooms, and databases help support a team's operations. This allows team members to participate in the team process whenever and wherever the opportunity arises.

Characteristics of Communication Technologies

The characteristics of communication methods can be used to analyze the differences among communication technologies (Table 15.2). Axley (1996) uses the following four criteria for evaluating communication methods: speed, reach (number of employees receiving the communication), interactivity, and cue variety (or richness). Reichwald and Goecke (1994) believe that social presence and media richness are the main variables for analyzing communication technologies. Richness relates to the speed of feedback, number and type of sensory channels, and how personal the source seems. Social presence refers to the degree to which using the technology resembles the experience of communicating with another person. This factor often relates to the richness of the communication media. Richness is important because people are more satisfied with richer communication technologies (Harris, 1993).

Another characteristic of many communication technologies is the ability to document the message or communication. These electronic records have important task and social impacts (Sproull & Kiesler, 1991). For example, the capability of communication technologies to document a message may inhibit managers from using them for fear of recording errors (Byrne, 1984).

TABLE 15.2

Analysis of Communication Methods

Method	Speed	Interactive	Richness	Social Presence	Document Message
Face-to-face	Slow	High	High	High	No
Group Meeting	Slow	Moderate	High	High	No
E-mail	Fast	Moderate	Low	Low	Yes
Web Page	Moderate	Low	Low	Low	Yes
Print	Moderate	Low	Low	Low	Yes
Video Broadcast	Fast	Moderate	High	Low	No
Videotape	Moderate	Low	Low	Low	No
Teleconference	Fast	Moderate	High	Moderate	No

SOURCE: Adapted from Levi, D., & Rinzel, L. (1998). Employee attitudes toward various communications technologies when used for communicating about organizational change. In P. Vink, E. Koningsveld, & S. Dhondt (Eds.), *Human factors in organizational design and management* (Vol. 6, pp. 483–488). Amsterdam: Elsevier Science.

It makes e-mail messages sent to team members valuable task reminders, given that the messages can be printed or stored.

An essential factor in analyzing the effectiveness of group communication is the richness of the communication. Richness of information media refers primarily to how much emotional information is transmitted (Daft & Lengel, 1986). The effectiveness of a communication technology depends on the fit between the task requirements and the richness of the technology. A task requiring the group to generate ideas requires only the transmission of the ideas, whereas negotiating or resolving a conflict requires providing the emotional contexts of the messages. The messages need to include the facts as well as the emotions, values, and expectations of participants.

Information richness has positive and negative effects. One reason that face-to-face groups are not as good as virtual groups for brainstorming is that the richness of the communication media gets in the way. In brainstorming, the presence of others and the social information being provided are a distraction from the task. It is inefficient to use a medium that is too rich. However, to use a medium that is not rich enough for a task is ineffective. The level of uncertainty in the communication increases the importance of information richness. As we will see, this is why technologies such as e-mail are not good for tasks such as negotiation.

2. Communication Impacts

When people communicate via technology, their communication is altered. Communication technologies change how status is perceived, the level of anonymity of the communicators, the conformity pressure from others, and the level of miscommunication. However, some of these changes are due to a lack of norms for managing the communication process, rather than to the impacts of the technologies.

Status Differences

One of the main differences between face-to-face groups and virtual groups is status differences. Research on student groups shows that status differences were reduced in virtual groups (Parks & Sanna, 1999). The reduction in status equalized participation in virtual group discussions. In face-to-face group discussions, a few dominant people with higher status in the organization talk the most; many group members limit their discussion and primarily support the main positions that emerge. In virtual decision making, the interaction is more democratic. Social cues are reduced, and people communicate on the basis of their knowledge or opinions rather than their social status.

Not all research on virtual teams has found this equalization effect. In a typical work group, higher-status members assume leadership positions, direct group activities, and are more likely to express their opinion (Driskell et al., 2003). Virtual teams may limit the effects of status because status cues are not as prevalent as in face-to-face communications. However, in actual organizational settings, group members are aware of the status of the other communicators regardless of the technology used. Field studies on existing teams with established status hierarchies (such as military or medical teams) show no communication change related to use of technology.

Status differences have a mixed effect on performance (Driskell et al., 2003). To the extent that status differences reflect differences in ability, expertise, or organizational knowledge, they may help a team's decision making. Reduced awareness of status differences may create uncertainty that increases stress in the team. Reduced status differences are more likely to affect judgment tasks where teams must reach consensus. Over time, the effects of status on a team's communications tend to decrease as members get to know one another better and communication becomes more linked to knowledge than to status.

Deindividuation

The members of virtual groups are more anonymous. This leads to what psychologists call deindividuation. Deindividuation is the loss of self-awareness and evaluation apprehension caused by feeling anonymous. It may have a number of negative social effects (Parks & Sanna, 1999). When not being evaluated, people are more likely to engage in social loafing in groups and are more likely to perform negative acts.

One of the impacts of deindividuation on virtual groups is that people are more willing to say things they would not say in face-to-face interactions. This is why "flaming," or uninhibited negative remarks, occurs in e-mail systems. This increase in uninhibited communication can be both positive and negative. When group members are anonymous, the communication becomes both more critical and more supportive. Basically, people feel freer to express emotions of all types.

Although flaming or other types of disinhibited emotional communication occur in laboratory studies, it is not a typical problem with less anonymous work teams (Hertel et al., 2005). Most virtual teams develop communication norms that regulate emotional communications.

Conformity

Pressure to conform is less in virtual groups, in part because of group anonymity. Lack of conformity pressure affects how groups manage conflict. Virtual groups have higher levels of conflict. Although conflict in any group may have both positive and negative effects, it often has negative effects on virtual groups. This is because the increased conflict is often due to increased misunderstanding of the emotional aspects of communication. In addition, virtual groups are less able to resolve conflicts and reach consensus in decision-making situations. These effects may be due to the lack of conformity pressure that would normally increase people's willingness to agree and seek compromises during disagreements.

Miscommunication

Virtual teams have increased conflict and miscommunication due to misunderstandings and reduced communication (Hertel et al., 2005). These communication problems increase the emotional frustration of people working in virtual teams. This is one reason why the effectiveness and member satisfaction of virtual teams is positively related to the level of personal communications among team members. Many of these problems go away as teams learn how to interpret communications from other team members.

Virtual teams have difficulty establishing and maintaining mutual knowledge (Driskell et al., 2003). They lack contextual cues during communications, leaving members less certain about with whom they are interacting, how the message is being conveyed, and whether the communication is successful. It is more difficult to know whether fellow members of virtual teams have adequately understood a communication. Facial cues and nonverbal feedback reduce the uncertainty one feels with a communication; without this information, inaccuracies and confusion occur that reduce team performance.

For example, the reduced social cues in e-mail messages make it difficult to communicate emotions, but people are often unaware of this problem and believe they are communicating effectively. E-mail readers often cannot tell if the writer is attempting to be sarcastic or funny (Kruger, Epley, Parker, & Ng, 2005). This overconfidence in the ability to communicate via e-mail is a cause of miscommunication in virtual teams.

Another essential cause of misunderstanding in virtual teams is the limited amount of communication that occurs (Driskell et al., 2003). One of the most consistent findings regarding the differences between virtual and face-to-face teams is that less information is communicated in virtual teams (Roch & Ayman, 2005). The increased information available to face-to-face teams is one reason why their decisions are often superior.

Communication Norms

Use of e-mail is a good example of how communication technologies require new communication norms or social interaction rules. E-mail has features that make it different from other types of organizational communication. Unlike the telephone or the written message, e-mail is asynchronous (i.e., the sender and receiver are not directly interacting) yet potentially interactive, private but capable of being distributed, and informal but can be used with technical precision (Culnan & Markus, 1987). The benefits of e-mail derive from these properties that are unique in the communication medium, but so do some of its problems (Kiesler, 1986).

Compared with other types of organizational communications, e-mail lies somewhere between the paper-based mail system and the telephone system. As with the mail system, it uses "printed" words that can be modified, stored, and distributed. As with the telephone, it is an instantaneous communication that links people across time zones and distance with ease. However, e-mail is more formal than a telephone call because it provides written documentation of the communication. In practice, e-mail is less formal than a written communication because it tends to be a shorter and more spontaneous communication that is less regulated by an organization's formal communication practices.

These differences between e-mail and other forms of communication have led to a number of "netiquette" errors caused by lack of established communication norms. Senders of e-mail messages often treat them like telephone calls (e.g., informal, private, no record), whereas receivers may treat them as letters (e.g., formal, public, copy, sendable). Senders are often shocked, and sometimes embarrassed, when their private e-mail messages are distributed to others. New users often show bad form by distributing "junk" news to people who do not care about it, writing overlong messages, or trying to be funny. Emotions are difficult for many people to put in writing (Kruger et al., 2005). In e-mail messages, emotions seem to get out of control easily, and interpersonal conflicts may arise that are primarily due to miscommunication. The solution is the development of communication norms for the technologies (Sproull & Kiesler, 1991). For example, a team may have a rule limiting the size of e-mail messages or does not allow using distribution lists for sending jokes.

3. Team Impacts

When teams depend on technology for communication, it changes how they perform. The impacts on team performance depend on the type of task. Decision making is affected, both directly and indirectly, by the use of communication technology. Many of these performance effects change as people become more experienced working in virtual teams.

Task Performance in Virtual Teams

From an individual's perspective, there are positive and negative aspects to working on a virtual team (Hertel et al., 2005). The advantages of virtual teams include flexibility, control over time, and empowerment of team members. Disadvantages include isolation, decreased interpersonal contact, and increased miscommunication and conflict.

Overall, the performance of face-to-face and virtual teams is fairly similar (Parks & Sanna, 1999). This is especially true if experienced users are being studied. Many of the disadvantages of virtual teams diminish or reverse when teams have had time to adjust to using the technology.

Although there are few overall differences in performance, some tasks may be better suited to virtual teams (Hertel et al., 2005). Virtual teams are more successful on idea generation and problem-solving tasks in which the team must organize information to find the correct answers. However, virtual teams perform poorly on decision-making and negotiation tasks in which the goal is to reach consensus on issues. For decision-making tasks, virtual teams take more time, exchange less information, and have lower member satisfaction than face-to-face teams. Most organizations prefer to use face-to-face communication for negotiation tasks because of the difficulty reaching consensus in virtual teams.

Again, brainstorming is the chief example of improved performance from virtual groups (Parks & Sanna, 1999). Virtual brainstorming groups achieve better ratings on all criteria. More ideas are generated, the ideas tend to be of higher quality, and people prefer to brainstorm using computers. Unlike face-to-face groups, the virtual brainstorming process is not disrupted by larger group size. The anonymity of virtual groups is a benefit in brainstorming because people are less concerned about criticism of their ideas. The only problem with virtual brainstorming is social loafing in large groups, but this occurs in large face-to-face groups as well.

Decision Making

The use of virtual groups has dramatic effects on decision making (Kiesler, Siegel, & McGuire, 1984). On the surface, virtual group decisions may appear to be superior because of the reduction of status cues that limit participation. However, there are both advantages and disadvantages to decision making in virtual groups.

During group decisions, members of virtual teams are more focused on the logic and factual basis of the arguments than on the social characteristics of the people making the arguments (Roch & Ayman, 2005). They are less likely to be distracted by irrelevant social information, such as gender, age, and race, that may lead to biased interpretations of the communication. This helps virtual team members to be better judges of the knowledge and abilities of other members. Virtual teams are better at weighing the quality of the opinions of individual members and folding this information into a group decision.

Although more people participate in virtual team discussions, there is mixed evidence that the quality of the decisions is superior (Watson, DeSanctis, & Poole, 1988). Members feel less committed to virtual team decisions, and higher-status employees may resent being unable to control the decision-making process. These factors may reduce the ability of teams to implement decisions. Finally, when issues are complex or many members are participating in the decision-making process, virtual team discussions may get so confusing that nothing gets resolved. Leadership, and perhaps limited participation, helps organize and manage the face-to-face decision-making process.

In most organizations, virtual teams play an important but limited role in organizational decision making (Levi & Slem, 1990). For complex and organizationally important decisions, most managers require face-to-face decision making. However, communication technology serves as a valuable resource in decision making, even when it is not directly used to make decisions. Organizations use technology to send more information to employees, so participants are better prepared when faced with making important decisions. The ease of using distribution lists means that employees not directly involved in a decision are at least aware of it. Overall, communication technology improves the quality of organizational decisions because of increased information and opportunities for participation.

Change and Development

People are less likely to be satisfied working in a virtual group than in a face-to-face group (Parks & Sanna, 1999). They may feel a lack of social support and experience increased stress working in a virtual team. Some of these negative effects are likely to disappear over time (McGrath & Hollingshead, 1994). Virtual work teams improve as they become more accustomed to using the technology, develop group norms for working together, and develop social relations that improve communication.

Virtual teams have a difficult time developing team cohesion (Driskell et al., 2003). Members of virtual teams are more anonymous; this leads to weaker relational ties and less team identification. Because of the limited ability to convey emotions via technology, it is more difficult for virtual teams to develop trust among members. This hurts the development of cohesion and makes it more difficult for teams to reach consensus.

Teams that have already developed good social relations may be more effective working in virtual teams. The social relations among members must be supported by occasional face-to-face meetings. When team members have never met personally, they are less effective using communication technology for difficult problems, because they have not developed the ability to "read" the emotional meanings of their mutual communications.

It is essential to take a lifecycle approach to understanding the impacts of using virtual teams (Hertel et al., 2005). The impact of technology varies depending on team stage. When new teams are created, a face-to-face meeting is important for team members to get acquainted with one another, clarify the team goals and member roles, and develop norms for operation. Virtual teams encounter a number of team maintenance problems. It may be more difficult to motivate virtual teams and control social loafing, create trust among members, and develop team cohesion and identification. Because of this, member satisfaction in virtual teams depends on the opportunities offered to meet face-to-face and exchange non-job-related, personal information.

Organizational Impacts

For organizations, virtual teams have a number of strategic advantages. The use of virtual teams allows organizations to organize teams of experts, have them work "around the clock," enable flexible response to customer and organizational demands, and reduce travel and office space expenses (Hertel et al., 2005). However, in addition to increased technology costs and technical problems, virtual teams create management difficulties in supervising employee behavior.

An advantage of virtual teams is the ability to select the best employees for the job, regardless of physical location (Hertel et al., 2005). Virtual teams often include members with differing expertise and cultural backgrounds, a diversity that creates communication problems and increases misunderstandings and conflicts, requiring that virtual teams have strong leaders. Because supervisors are unable to directly monitor employees, management approaches that delegate authority, such as management by objects or goal setting, are valuable for virtual teams.

4. Selecting the Right Technology

Virtual teams have a variety of available technologies to use. Teams may have problems because they fail to consider whether technologic features match organizational requirements (Caldwell & Uang, 1994). Selecting the right technologies requires consideration of both organization and team issues.

Organizational Communication

The selection of appropriate technologies for a virtual team depends on the task and user attitudes toward the communication and communicators (Reichwald & Goecke, 1994). Too often, communication specialists focus on technologic capabilities rather than on understanding users' information needs and the effects the medium will have on users (Gallaway, 1996).

The more powerful (or rich) multimedia communication technologies might not be the best for all tasks (Dertouzos, 1997). For example, because contracts require specific wording and the ability to store and study documents, print may always be the preferred communication medium. Multimedia communication can be distracting to users. Adding audio to a virtual team may improve communications by better conveying the social and emotional aspects of communication (Driskell et al., 2003). However, there does not seem to be an added benefit to videoconferencing over audioconferencing. In some cases, professionals turn off the pictures during videoconferences in order to concentrate on the messages.

The effectiveness of a communication technology depends on user attitudes toward the communication. The level of uncertainty of a message and the level of impact on the receiver determine which communication technology will be most appropriate (Axley, 1996). The greater the uncertainty and impact, the more the message needs to be interactive, more broad in reach, and richer.

Preferences in communication technology depend on the attitude toward the communicator (Levi & Rinzel, 1998). When the source of the communication is seen as trustworthy, nearly any medium is acceptable. When it is not, however, face-to-face or other information-rich communication is required to evaluate the truthfulness of the message.

Differences Among Types of Teams

Different types of teams have different communication needs, so the right information technology depends on the type of team and its tasks (Bikson, Cohen, & Mankin, 1999; Mittleman & Briggs, 1999). Production teams use information technology to coordinate activities, track work progress, communicate with other sectors of the organization, monitor processes, and analyze production information. Information systems provide critical data and help team members analyze and use the information.

Members of service teams may be widely distributed across space, so they are dependent on communication technologies and shared databases. Because providing services is often time-dependent, information technology helps service team members to coordinate their activities. Coordination is improved through better communication and the ability to share databases containing information about customers and team members.

Technology allows project teams to form without regard to place, allowing virtual teams to include members with crucial skills or information who could not otherwise be on the team because of work location. Dispersed project teams depend not only on communication technologies, but also on shared databases all members can contribute to and access. In addition, project teams are more likely to use computer support systems to help them perform their work and manage the team process.

Management teams tend to make heavier use of videoconferencing than do other types of users. This reflects their preference for face-to-face rather than written communication. They prefer technologies that are easy to learn and have intuitive user interfaces and do not document all communications.

Team Communication

There is disagreement whether virtual teams create their own new social contexts, or simply transfer their existing social contexts into the new situations (Parks & Sanna, 1999). There are a number of similarities in how face-to-face and virtual groups operate. There are differences, but the basic group processes appear to be the same.

Virtual teams adapt and modify communication technologies to fit their needs; teams select different technologies for different types of team tasks. E-mail is used to keep team members informed and share information, but rarely for negotiation and decision making. Videoconferencing is useful for discussing issues and listening to presentations, but when teams need to work on technical issues, they often prefer audioconferencing with shared computers. This allows them to focus on tasks without the distraction of video images of members. Project management software is used so team members can monitor one another's performances and manage projects as teams.

There may be limits to the ability of virtual teams to fully adapt. Virtual teams are highly task-focused. This is useful in the short run, but virtual teams do not develop social relations. Their lack of social development creates communication problems. People do not learn how to "read" the emotions in each other's communication and do not develop a sense of mutual trust. Such problems are often managed by having teams meet face-to-face for initial and periodic review meetings. Face-to-face meetings may not be necessary for task purposes, but they are necessary for the social development of the team. Without social development, teams may become mired in conflict and failed negotiations through miscommunication.

Summary

The use of communication technologies has expanded rapidly and is changing how teams operate. The new technologies increase access to information, support internal and external communication, and help teams manage their task and group processes. Virtual team meetings are no longer constrained by time or place. Differences among communication technologies are related to their speed, interactivity, richness, social presence, and ability to provide documentation of communications.

Communicating through the technologies has interpersonal effects. Status differences are reduced, and interactions are spread more evenly among participants. People are more anonymous in virtual groups, but this can increase negative or emotional messages. Conformity pressure is reduced, making it more difficult for groups to manage conflicts. Reduction in social and emotional cues increases the chances for miscommunication. Many of these effects reflect the absence or insufficiency of communication norms for regulating behavior.

Virtual teams differ from face-to-face teams. The former perform better on idea generation and problem-solving tasks but worse on decision-making and negotiation tasks. Decision making is improved in virtual teams by more equal-status communication. However, virtual teams may be less likely to develop support for implementing decisions, and these teams have a more difficult time managing conflict. Virtual teams are better prepared to make decisions because of the increased availability of information. One of the greatest problems for virtual teams is developing social relations with members.

Selecting the right communication technology depends on the characteristics of the technology, users, and task. The most advanced technology may not be the best, given that rich technologies can be distracting. Over time, teams adapt to communication technologies through modification of technologies and the development of new communication norms and procedures.

TEAM LEADER'S CHALLENGE—15

You are the leader of a virtual team that is coordinating research projects among your corporation's five research centers, which are distributed around the world. Although you had a coordinators' meeting several years ago, cost and time constraints make meeting regularly impossible. The research centers have videoconferencing equipment, but time differences among the sites make using videoconferencing services difficult. Consequently, most of your team's communication is done via e-mail.

The virtual team has worked well at exchanging information and keeping everyone up-to-date on the progress of research. However, there is a growing conflict between one of the U.S. research centers and the Asian center; they seem to be unable to coordinate activities and negotiate project roles. Their e-mails are getting more critical and disrespectful, and the rest of the team is tired of reading their back-and-forth bickering.

How can the team leader deal with this communication problem?

Does the solution require either face-to-face or videoconference meetings?

Could establishing new team e-mail norms be used to prevent such problems in the future?

ACTIVITY: DEVELOPING NETIQUETTE FOR VIRTUAL TEAMS

Objective: Virtual teams need to develop different norms to regulate team interactions. Norms for virtual teams are sometimes called netiquette rules.

Activity. Develop a set of norms for a virtual team (Activity Worksheet 15.1). These norms should cover communication, participation, and decision making. What other types of norms are needed to help a virtual team operate effectively?

ACTIVITY WORKSHEET 15.1

Norms for Virtual Teams

Communication Norms:
Participation Norms:
Decision-Making Norms:
Other Team Norms:

Analysis. Compare the norms you have developed with those of other members of the group. Through discussion, develop a common set of norms for virtual teams.

Discussion. What are the advantages and disadvantages of virtual teams? Can the development of new norms improve how virtual teams operate? What other actions should be taken to support the use of virtual teams?

16

Teams at Work

Increased use of teams at work has led to major changes in how organizations operate. This shift to teamwork provides a variety of benefits and challenges to organizations. The effectiveness of teams depends on the characteristics of the organization, the type of team, and the nature of the workers. Examples from production, professional, and managerial teams illustrate the different advantages and disadvantages of teamwork. Regardless of type of teams, there are supportive actions that organizations need to perform to resolve common problems that teams encounter.

Learning Objectives

1. What are the main benefits of using work teams?

2. What problems do organizations encounter when using teams?

3. What are the implications of teams being a fad?

4. Understand the risks and opportunities in using different types of teams at work.

5. What are the main features of factory work teams?

6. What characteristics make professional teams unique?

7. Why is using top management teams difficult?

8. What types of supports do organizations need to provide for work teams?

9. What are the common problems organizations encounter when using teams?

1. Using Teams in the Workplace

Work teams are viewed as an important way of improving organizational effectiveness. The transition to use of work teams in factories and offices is considered necessary to help corporations remain competitive (Gwynne, 1990). In fact, the implementation of work teams is one of the most common organizational interventions in manufacturing firms (Sundstrom, DeMeuse, & Futrell, 1990). It is one of the most effective interventions for improving organizational performance (Guzzo & Dickson, 1996). In addition to increasing the financial success of companies, teamwork programs improve personnel issues such as turnover and absenteeism.

Use of work teams is growing in the United States. In a 1987 study of Fortune 1000 companies, it was estimated that 70% of the companies used some type of work teams or problem-solving groups, whereas 27% of the companies used self-managing teams (Lawler, Mohrman, & Ledford, 1995). When the study was replicated in 1993, 91% of the companies reported using work teams and 68% were using self-managing teams. Although the use of teams is widespread, most employees do not work in teams. In most cases, fewer than 20% of a company's workers are members of teams.

Most of the emphasis on teamwork during the 1980s involved production or service employees (Safizadeh, 1991). Teamwork was seen as a new way to organize work, with decision making shifting to the employees actually performing the tasks. Since the early 1990s, the focus of teamwork activities has changed. On the factory floor, companies that have been successful in developing teams are continuing their change efforts by increasing their self-managing teams (Manz, 1992). Teamwork programs have expanded through the organizational hierarchy to include teams of professional and managerial employees (Katzenbach & Smith, 1993).

Benefits of Teamwork

Teamwork is increasing because teams are an effective way to improve performance and job satisfaction. Large-scale studies on the use of production work teams show their effectiveness (Guzzo & Dickson, 1996). Teams improve both the efficiency and the quality of organizational performance. Using teams provides the flexibility needed to operate in today's rapidly changing business world. When work teams are widespread in an organization, the organization tends to show improvement in other performance areas, such as employee relations. However, teams may develop performance problems that limit their effectiveness, and the initial transition to teamwork may be a difficult process for organizations.

In addition to increasing organizational effectiveness, the implementation of work teams often leads to improvements in job satisfaction and quality of work life (Sundstrom et al., 1990). Teams have these beneficial characteristics because they provide social support to employees, encourage cooperation, and make jobs more interesting and challenging. In addition, the transition to teamwork requires training that improves employees' technical and interpersonal skills. This additional training is viewed as a personal benefit by employees.

The benefits of teamwork are realized only when teams are working on jobs that are suited for teamwork and organizations are willing to support them. Table 16.1 presents a set of task and organizational characteristics that are necessary conditions for the use of teams.

Problems of Teamwork

Although there are benefits to both organizations and employees, some problems are created by the use of teams and the transition to teamwork.

TABLE 16.1

When Are Teams Appropriate?

1. The work contains at least some skilled activities.

2. The team can form a meaningful unit with the organization, with clearly defined input and output and stable boundaries.

3. Turnover in the team is minimal.

4. Valid performance evaluation systems exist for both the team and its members.

5. Timely feedback is possible.

6. The team is capable of measuring and controlling the important variances in the workflow.

7. The tasks are highly interdependent so members must work together.

8. Cross-training is supported by management.

9. Jobs can be designed to balance group and individual tasks.

SOURCE: Adapted from Davis, L., & Wacker, G. (1987). Job design. In G. Salvendy (Ed.), *Handbook of human factors.*

Research on teamwork in work settings provides mixed results. Many of the studies on quality circles (i.e., temporary teams that provide suggestions about how to improve quality) show that these teams are not effective, whereas studies of factory work groups have widely variable results (Guzzo & Dickson, 1996). One of the problems is that teamwork programs are implemented with little consideration for their applicability, and rather than attempting to make existing programs work better, new programs are introduced.

Teamwork programs such as quality circles provide only limited power to teams. Such programs often lead to small short-run improvements in performance, but not to long-term improvements (Guzzo & Dickson, 1996). The shift to self-managing work teams often results in significant long-term performance improvements. However, the transition to self-managing teams can be difficult in organizations with traditional management control systems.

Effective work teams have norms that support high-quality performance and a level of group cohesiveness that provides social support for members. However, work teams may have problems with norms and cohesiveness. Teams with poor performance norms may not be effective and may be highly resistant to change. Low levels of group cohesion may limit a team's ability to work together, whereas high levels of group cohesion may lower members' performance orientation and impair decision making (Nemeth & Staw, 1989).

Implementing work teams often creates problems. Conflicts exist between team development and the traditional management systems in many organizations (Hackman, 1990a). Teams suffer from implementation problems due to resistance to change. Teamwork requires a supportive organizational context to foster team growth and development.

Teams Are a Fad

Many managers and employees overrate the effectiveness of teamwork (Allen & Hecht, 2004). People overemphasize the success of work teams because of the psychological benefits of teamwork, which make them assume that teams are high performing. Although research on the effectiveness of work teams is mixed, these mixed findings are not synchronous with the positive view of teamwork held by many managers. The implementation of work teams has been one of the most common organizational changes of the last 20 years. Teamwork has become a business fad and is suffering from problems of overuse (Charan & Useem, 2002).

It is difficult to evaluate the effectiveness of work teams because the implementation of teams often occurs along with other organizational changes (Paulus, 2002). There is usually improvement when teams are first introduced, but the positive effects of teamwork fade (Allen & Hecht, 2004). Part of the problem is that because teams are a fad, they are overused. In certain situations, work teams are an appropriate task structure, especially when jobs are interdependent, but this does not mean that teams are the solution for all organizational problems.

The team "halo effect" is one reason why people overemphasize the value of teams (Naquin & Tynan, 2003). Success in teams is attributed to teamwork; lack of success is attributed to poor individual performers or outside factors. Teams also provide their members with psychological benefits that lead members to overemphasize their value (Allen & Hecht, 2004). Being on a team satisfies social needs at work; this is especially important in large, impersonal organizations. Because people enjoy working with others, teamwork improves their attitude toward their work. Teams provide members with a social identity that reduces stress and uncertainty. Members of teams have the illusion that their team is high performing, partially because of the social support they receive from the team.

One result of the overly positive view of teamwork is that use of teams has expanded beyond where they are valuable (Allen & Hecht, 2004). They are used to solve every organizational problem, regardless of whether they are an appropriate way to organize the work. This means that many teams operate in organizational contexts that are inappropriate for teamwork. The strong belief in the effectiveness of teams also leads to implementation of teams without the organizational changes needed to support them (Charan & Useem, 2002). Managers implement teams, looking for benefits without considering the costs of training teams and other organizational changes (Paulus, 2002).

What is needed is a better understanding of where and when teams should be deployed and what actions are required to deploy them effectively. Teams are not the solution to every organizational problem, and they are not automatically successful. They need a purpose, an outcome that requires joint efforts, complementary skills, and mutual responsibility (Katzenbach & Smith, 2001). Organizations get in trouble when the goal of a team is promoting teamwork (a process goal) rather than an identified performance outcome.

2. Differences Among Work Teams

Teams exist at different levels of the organizational hierarchy, ranging from production to professional and managerial teams. At each level, the use of teams involves different risks and opportunities for their organizations. These are outlined in Table 16.2. In the following sections, production, professional, and managerial teams are examined in more detail.

TABLE 16.2

Risks and Opportunities for Work Teams

	Risks	*Opportunities*
Production Teams	• Controlled by technology • Resistance from management	• Continuity of work • Ability to improve team and product
Professional Teams	• Team and work are both new • Evaluation and rewards	• Clear purpose and deadline • Interdependence
Top Management Teams	• Absence of organizational context • Competitive relationship	• Self-designing • Substantial power

SOURCE: Adapted from Hackman, R. (1990). Creating more effective work groups in organizations. In R. Hackman (Ed.), *Groups that work (and those that don't)*. San Francisco: Jossey-Bass.

Production teams run the risk of being overwhelmed by their tasks and the technology used to perform them. The technology controls how these teams operate, leaving little leeway for them to make changes. Because of their focus on production, members may ignore the social development of their teams. Supervisors may resist giving production teams the responsibility and authority they need to operate as teams. However, team members have the advantage of working together for long periods of time, which gives them the opportunity to refine both their teams and the tasks they perform.

The main problem for professional teams is that they must perform unique jobs for which they cannot fully prepare in advance. Their projects are typically time-limited so they must swiftly develop structures and processes to operate. The main advantage of teams is that they often have specific goals and deadlines. In addition, their tasks are highly interdependent. These factors orient and motivate team members to work together.

Top management teams suffer in the absence of supportive organizational contexts. They operate in a competitive environment that is not supportive of teamwork. The special opportunity for top management teams comes from their power in their organizations; their actions have significant impact on the operations of their organizations. In addition, such teams have the ability to set their own goals, thereby allowing them to focus on the issues they view as most important for their organizations.

Production Teams

The increasing competitive pressure of the 1980s forced U.S. companies to search for alternatives ways of operating. One of the main approaches they adopted to meet this challenge was the use of work teams. Three factors predict the adoption of work teams by manufacturing companies: competition internationally, high-skill technology, and a worker-oriented corporate culture (Appelbaum & Batt, 1994).

There are two major types of factory teams: teams that are integrated into jobs and teams that operate as parallel structures (Appelbaum & Batt, 1994). In the integrated approach, jobs are designed to be performed by teams. Workers are then given the power and authority to continue redesigning their jobs to promote continuous improvement in operations. In the parallel approach, the jobs remain the same, but workers are organized into teams in order to analyze work issues to improve quality or performance. These teams have specific functions that are distinct from members' work assignments. The main problem these teams encounter is that the solutions they generate might not be implemented by their organizations.

Factory work teams vary in levels of power and areas of control. From a leadership perspective, teams range from supervised to self-managing teams. Most companies use supervised teams, but self-managing factory teams are increasing. Work teams also vary widely in their areas of control. Some teams have control only over direct work activities, whereas other teams control personnel issues such as hiring, training, and performance evaluation.

A new form of factory teamwork emerged as a major organizational form in the United States during the 1980s (Appelbaum & Batt, 1994). It combined aspects of the European sociotechnical systems approach and the Japanese lean production model with American human resources policies. This model, called American team production, leads to a redistribution of power in factories so that work teams have the ability to control their own operations. Possibly the most famous example of this new model is General Motors' Saturn facility. The main features of the model relate to the organization of work, human resources policies, and quality.

In American team production, work is reorganized into self-managing teams. Continual improvement is expected, given that workers have decision-making authority over their jobs. It is up to the teams to establish and modify their work processes as needed. These teams vary in their areas of control and their independence from management. At a minimum, they control their own work processes and work areas.

The team structure creates the potential to improve performance, but it is changes in human resources policy that provide the incentive. Teams are often given control over hiring and training new employees, as well as authority to conduct performance evaluations and other personnel functions. They typically work under team incentive programs that reward improvement in team operations.

In most cases, quality is a central concern of factory teams. If only cost and productivity were important, the time lost in team meetings might not be worth the expense. The benefits of teamwork lie not in short-run cost savings, but in quality and long-run performance improvement. Without concern about quality and the potential loss of markets, managers would rarely be willing to support the investment required in developing teams.

The shift to work teams in factories faces a number of barriers (Appelbaum & Batt, 1994). These relate to the existing technical infrastructure, unions, and the managerial power structure. Often production technology is designed for individuals operating separate machines (Hackman, 1990a). This is an impediment to the use of teams, which is why some of the best examples of factory teamwork are in new factories designed for teams. Although unions were initially hesitant to support such changes, the workers' positive attitudes toward teams have made unions supportive of them. The greatest impediments lie in management. Managers are not accustomed to sharing power with work teams. That the shift to teamwork often leads to reductions on organizational levels (therefore, reductions in the numbers of managers) also discourages managerial support of teams.

Professional Teams

Professionals are often organized into task forces or project teams. These are temporary teams brought together to perform single tasks, after which they are disbanded. There are several unique characteristics of these types of professional teams. Their tasks, group processes, products, and team members are unique and unrepeatable because the teams are formed to perform tasks and then are disbanded. Their projects are usually given limited time frames, so they must go quickly through all the stages of group development.

Project teams are typically composed of members from multiple functional areas in organizations, with little overlap of skills and knowledge (Gersick & Davis-Sacks, 1990). Consequently, the team's ability to discuss, negotiate, and make group decisions is vital because there are no authorities with the ability to make better decisions.

The nonroutine nature of the tasks and the short-term relationships among team members make developing teamwork norms and procedures a stronger priority than for other teams. These teams do not have the luxury of allowing group norms to evolve slowly while the team develops. Team development is more difficult because members do not know what their projects will require at the onset.

Project team members face unique conflicts. They are both autonomous and dependent. They may act with relative independence because the project boundaries are distinct from the normal actions of their organizations. However, all team members answer to both their team and the division of the organization that they represent. When the project is over, members return to their functional areas. This creates a loyalty conflict in individual team members.

One important type of professional project team is the research and development (R&D) team. R&D teams have characteristics that may make them operate differently from other teams (Edosomwan, 1989). The personnel on R&D projects are mainly technical professionals with highly specialized knowledge backgrounds. Most organizations separate their R&D teams from operations personnel and manage them differently. The tasks performed in R&D are by definition nonroutine; therefore, traditional management approaches that focus on control are inappropriate. The unique characteristics of R&D teams have both advantages and disadvantages for teamwork development. One of their main advantages arises from the nature of their task (Hackman, 1990a). In most cases, R&D projects require the integrated efforts of professionals with a variety of technical skills. The complexity of the task requires multiple skills and mutual interdependence. Although this can sometimes be accomplished without teamwork, the team approach is an obvious benefit. In addition, R&D work is challenging, important, and viewed as high status compared with other professional work; this promotes a sense of teamwork.

R&D teams have some significant disadvantages in developing and supporting teamwork. R&D professionals are selected for their professional skills, which do not necessarily include social and communication skills (Carnevale, Gainer, & Schulz, 1990). R&D professionals tend to be individualists who would prefer to be rewarded for personal effort (Ellis & Honig-Haftel, 1992). This discourages their commitment to teamwork, where their individual efforts are not easy to identify and reward. Because R&D projects are complex and ill-defined, it is often difficult to determine whether success or failure is due to member efforts or to the nature of the task (Sundstrom et al., 1990). In many organizations, the matrix management structure, in which technical professionals have both project and functional managers, further complicates performance evaluation (O'Dell, 1989). These dual loyalties may reduce commitment to the project team.

Top Management Teams

Forming top management teams is difficult, and it is not clear whether such teams are needed to run organizations (Katzenbach & Smith, 1993). Top-level managers are often reluctant to work in teams. However, when top management teams are used to set strategic directions for organizations, their decisions have great impact. Compared with other types of teams, these have considerable power, authority, and independence.

Most organizations are run by working groups rather than by top management teams. In these groups, there are strong leaders and individual work products and accountability. Although these groups discuss issues and make decisions, the managers who are members operate independently. The groups provide advice to the leaders, but they do not perform integrated tasks for which members are mutually accountable; they are not a team.

There are several reasons why an organization would shift from a single leader to a top management team approach (Eisenstat & Cohen, 1990). A team's decision is more likely to represent the wide variety of interests in the organization. A team with members of varied backgrounds is better able to develop creative solutions to problems. Participation in the decision-making process is likely to encourage more support for implementation of decisions. Communication and coordination among the major divisions of the organization should improve with teamwork at the top. Finally, participation in a top management team develops the skills of its members and makes the organization less vulnerable to disruption from turnover at the top.

One of the most difficult problems for top management teams is managing the level of competition among the team members (Eisenstat & Cohen, 1990). The ultimate responsibility for the success of an organization typically rests with the leader. In U.S. culture, it is difficult to develop the notion of mutual accountability and responsibility at the top levels of an organization. Even for organizations with top management teams, the notion of single organizational leaders is difficult to avoid. This is the main barrier to developing successful teams at the top.

Being members of a top management team is significantly beneficial for individual participants (Eisenstat & Cohen, 1990). Not only does it improve their careers, but it is valuable training in running an organization. However, other managers not on the team may resent not being included. Use of top management teams may create power struggles among upper-level managers. Consequently, some team members play it safe and try to agree with the leader. This type of conformity can reduce the value of the team.

3. Supporting Work Teams

Work teams require several types of organizational support to be effective (Sundstrom, 1999b). These organizational supports are: establishing an appropriate foundation for teams, supporting the team process (leading, training, measuring and providing feedback, and rewarding teams), and establishing a proper infrastructure.

The foundations of teamwork relate to establishing a team structure and staffing the team. The organization needs to establish an appropriate structure for the work team by defining the team's responsibilities and authority, establishing the team's scope or boundaries, providing sufficient resources, and developing a system of accountability. Selection and staffing issues ensure that the team has the right mix of knowledge, skills, and abilities to complete the task.

The main team processes are leading, training, providing feedback, and rewarding the team. The team leader must coordinate the team's actions with the rest of the organization, provide direction for the team, and support team members through coaching. Training is needed to develop teamwork skills. Performance measurement and feedback systems improve how the team functions. Reward systems provide incentives for individual and team performance and encourage cooperation among teams in the organization.

A team's infrastructure relates to the availability of information resources, communication technology, and the design of the physical work environment. Information systems should provide the team with responsive data access for supporting the task. Communication technology links team members and provides linkages to distant team members, customers, and suppliers. The physical environment supports teamwork by providing space for individual and team work and informal social interaction.

The importance of these various support systems depends on the type of team. Measurement and feedback systems are essential for production and service teams, which need to track performance to improve how they operate. Communication and information technologies are important for project teams because of the analytic nature of their work and the need to coordinate members who may reside in different locations.

4. Problems With Using Work Teams

Although organizations use a variety of team types, there are important commonalties among teams at work. Hackman (1990a), in an analysis of a variety of work teams, identified several "trip wires" or common problems in the design and management of work teams. Reviewing these trip wires illustrates the challenges that must be surmounted for teams to be successful.

Calling People a Team but Managing Them as Individuals

Employees can be organized to work in one of two ways. Job assignments are given to individuals, with a supervisor controlling and coordinating the activities of the whole. Alternately, the task is handed over to a team whose members have joint responsibility for performing and managing the task. Either way can be successful. What is not successful is treating people as individual performers and telling them they are a team. This not only confuses employees, but creates cynicism about teamwork.

There are several reasons why this problem occurs. The notion of teams is trendy; sometimes managers refer to workers as teams for this reason. In other cases, organizations try to set up teams, but managers do not want to give the teams sufficient power and control. The organizational culture may be so individualist that teams do not fit into the environment. When the reward system is individually oriented, it is difficult to get employees to commit to teams.

Teams must be designed to benefit organizations. This includes establishing boundaries for the teams, defining group tasks with collective responsibility, and giving teams authority to manage these tasks. In some organizations, this is impossible because the organizational culture does not allow it.

Not Providing the Right Amount of Authority

Managers and teams must balance the issues of authority and responsibility in organizations. Teams can be given too much authority, especially if they are not mature enough to know how to operate. Managers may be concerned about handing over authority to teams when their organization still holds them responsible. However, if managers try to retain too much control over teams, most of the benefits of teamwork are lost.

Achieving a good balance between a team and management is difficult. There is a question not only of how much authority the team should have, but also which areas of responsibility should be covered. A team is able to handle more authority and responsibility as it matures. The difficulty for a manager is to not give away too much authority at the beginning of the team's life, and to learn when to turn over more authority as the team gains experience.

Providing a team with a clear goal empowers the team. A clear direction helps it form its own objectives and motivates performance. Without clear direction, a team will often become mired in unproductive activity. Once the goals are established, the organization should turn over to the team the authority to control how the task will be accomplished.

Assembling a Group of People, Giving Them General Goals, and Letting Them Figure It Out

Often the jobs individuals perform are overly defined. Employees are locked into ways of working that may not be optimal. When work teams are formed, they are viewed as a way of freeing employees from past constraints. However, teams are not effective if they are simply told to figure out what their jobs are; this gives them less structure than they need to operate successfully.

Teams require structure that defines their tasks and membership. The limits of their authority should be explicitly stated. When they are given appropriate structures, teams are able to focus on developing their internal processes for performing their tasks. When they are not given appropriate structures, teams become mired in unfocused attempts at creating them.

An enabling structure for a team has three components. First, a well-designed team motivates its members through meaningful work, enough autonomy to perform the task, and feedback about results. Second, a well-structured team has clear boundaries and includes members with sufficient skills and knowledge to perform the task successfully. Finally, such a team has a clear understanding of the extent and limits of its authority and accountability.

Specifying Challenging Team Objectives Without Providing Adequate Support

A challenging objective can provide a team with direction and motivation, but without sufficient resources to perform the task, the team will eventually stop being motivated by the goal. For the team's full potential to be realized, the organization must actively support teamwork. The types of support required by a team include a reward system that recognizes team performance, an educational system that provides training and technical support, an information system that helps the team make decisions and monitor performance, and the material resources to complete the task.

It may be difficult for an organization to supply these supports, especially if the organizational culture has an individualist focus. It is difficult to change performance evaluation and reward systems. The organization may not have trainers to support team development. Information systems may be designed for managers and not easily altered to support a team. Finally, it may be difficult for a team to adapt the existing technology and space for its needs or to obtain resources from the organization.

A team-oriented organization differs from a traditional organization, and the transition from one to the other is not an easy one to make. The work systems designed to support and control individual work are often not amenable to change. In addition, not all managers want to support an organizational innovation that threatens their power.

Assuming That Members Already Have the Skills and Knowledge They Need to Work as a Team

Once teams have started, organizations sometimes just leave them alone. There are good reasons not to interfere too much with the internal operations of a team. However, a hands-off approach can limit the team's effectiveness if its members do not have the skills they need.

One valuable role of managers is providing coaching for teams to help develop member skills in teamwork. There is no one best way to lead work teams; the needs of different teams and maturity levels of members are too variable. However, coaching is not a substitute for structure, clear goals, and the resources to succeed. When a team is failing for lack of these contextual factors, coaching will not help.

Finally, what happens early in a team's existence is likely to have continuing effects. Because of this, it is important for the leader to help the team get a good start. Building the team at the beginning prepares it to handle problems and crises later. The leader should not wait until a crisis occurs before teaching the team how to manage its group processes.

Summary

The use of teams in the workplace has been increasing during the past several decades. This trend started with factory teams and moved up the organizational hierarchy. Teams are popular because they are effective at improving performance and job satisfaction. However, they are not useful in all situations. Teams run into problems when they are not used appropriately, and they often encounter resistance in nonsupportive organizational environments. Unfortunately, the popularity of work teams has led to their overuse in some organizations.

Different types of work teams have their own sets of risks and opportunities. Production teams are ongoing, giving them the time to develop as teams, but they are often controlled by outside forces. Professional teams have clear purposes and deadlines, but must form anew for each project. Top management teams have the power to implement decisions, but their members have difficulty working together cooperatively.

Factory teams can either be integrated into their jobs or used in parallel to promote improvements in quality or other factors. Factory teams vary in power and areas of control. A new model of factory teamwork has developed that uses self-managing teams, supportive human resources policies, and an emphasis on quality. Although factory teams face a number of barriers to implementation, they have a track record of success.

Professional teams are often temporary teams organized around projects. The unique nature of the tasks and the lack of existing relationships among members make forming effective teams challenging. R&D teams are an example of professional teamwork. These teams have the benefits of motivating tasks, and they receive high levels of support from their organizations. However, they may lack people with teamwork skills and the right organizational incentives for working together.

Top management teams have the greatest potential to affect the success of organizations but may be the most challenging type of teams to develop. Single leaders run most organizations, and it is often unclear what the roles of top management teams should be. In addition, upper-level managers tend to be very competitive, making communication and coordination potentially difficult. However, these are highly competent people with a diversity of skills and knowledge, whose potential to develop and implement creative solutions to problems is great.

All types of teams require support from their organizations to operate successfully and meet challenges in their development. The organizational support required for a team includes establishing effective team structures; supporting the team process with leadership, training, and rewards; and providing appropriate physical and technological infrastructures. The challenges organizations encounter in using teams are related to power and authority, development of appropriate jobs and goals, and provision of resources and leadership.

TEAM LEADER'S CHALLENGE—16

You didn't think reorganizing a factory for teamwork would be this difficult. The original teamwork meetings on the factory floor made you aware of language problems among the employees, especially the immigrant workers. You thought everyone would welcome cross-training and learning new skills, but some older workers have told you they prefer doing their current job and don't want to learn new skills. The supervisors are having a hard time acting as team coaches because they are more comfortable with their former style of managing employees. There have also been rumors that once the teamwork program has been completed, the need for supervisors will drop. In the new system, the factory teams are to interact directly with the technical professional staff. However, the technical professionals are more comfortable interacting with the supervisors and are resisting interacting directly with the factory teams. Upper management expects the teamwork reorganization to be completed in a brief period of time, but you have so many problems that you are not sure where to begin.

What should be your focus for promoting teamwork among the factory workers?

How should resistance from the supervisors be handled?

How can you improve relations between the factory teams and the technical professionals?

How can you make management's expectations more realistic?

ACTIVITY: UNDERSTANDING THE DIFFERENCES AMONG WORK TEAMS

Objective. Work teams exist at different organizational levels. Production and service teams typically are used to produce products or provide services, professional teams are used to plan and coordinate projects or design new products, and management teams are used to run organizations and plan organizational change programs. These teams perform different functions and must face different issues to succeed.

Activity. For each type of work team (Activity Worksheet 16.1), analyze and discuss the biggest obstacles the team faces, the biggest benefit to the organization of using this type of team, and the type of support the team needs from its organization.

ACTIVITY WORKSHEET 16.1

Analysis of Work Teams

Production & Service Teams	Obstacle:
	Benefit:
	Support:
Professional Teams	Obstacle:
	Benefit:
	Support:
Management Teams	Obstacle:
	Benefit:
	Support:

Analysis. Compare the three types of teams, noting their commonalties and differences.

Discussion. Given the information derived from this analysis, what should organizations do to encourage the successful use of teams in the workplace?

17

Team-Building and Team Training

Team-building is the term used to describe approaches to improving the operation of teams. Team-building programs typically focus on improving teamwork skills, developing social relations, and solving problems that disrupt team performance. Organizational training programs may also be used to develop more effective teams. Training programs teach specific teamwork skills, such as assertiveness; cross-train team members to improve coordination; and promote organizational learning through the use of problem-solving teams.

Learning Objectives

1. Understand some of the different definitions of team-building.

2. What are some reasons why organizations do not use team-building?

3. What are the criteria of effective teams and the symptoms of ineffective teams?

4. What are the main types of team-building activities?

5. What skills are taught in teamwork training programs?

6. What factors relate to the effectiveness of teamwork training?

7. What are the main types of teamwork training?

8. How does action learning develop and improve teams in an organization?

1. What Is Team-Building?

Organizational development is a set of social science techniques designed
to promote changes in organizations that enhance personal development and
increase organizational effectiveness. Team-building is an organizational development intervention that focuses on improving the operations of work teams.
Team-building means regularly taking time to evaluate the performance of
teams to identify obstacles and develop more productive patterns of work. To
be effective, team-building must be viewed as an ongoing activity.

A variety of perspectives on team-building emphasizes different goals and
approaches. Rather than trying to combine these perspectives to create a single
view, the following is a sample of the different perspectives on team-building:

- Team-building is a problem-solving process that focuses on the following three
  issues. What keeps the team from being effective? What changes could improve
  the team's effectiveness? What is the team doing effectively now that it wants
  to continue doing (Dyer, 1995)?
- A team-building program uses three tactics. First, the program may have a task
  focus that examines the team's problems and attempts to develop solutions to
  them. Second, it may take a group process or relationship focus, using exercises and activities to improve how the group operates and the interpersonal
  relationships among team members. Third, it may take a structural approach
  and develop new norms, rules, and procedures to improve the team's operation (French & Bell, 1984).
- Team-building means making sure the team has common goals and that
  members can work together to achieve them. The main priority in team-building
  is developing a strong sense of belonging to the team. Unless team members
  identify with one another and see themselves as a team, it will be impossible to
  organize them to accomplishing a common goal (Hayes, 1997).
- What does it mean to develop a team? It means creating a team with the appropriate mix of skills, including technical and group process skills. It also means
  improving performance by changing the way the team operates. The team
  development process includes organizing work and roles, acquiring the necessary skills and resources, establishing the necessary relationships inside and outside the team, and changing the situation to facilitate performance (Mohrman,
  Cohen, & Mohrman, 1995).
- Team-building assumes that a successful team can be developed through training
  and practice. A team must learn how to set goals, structure work assignments,
  coordinate efforts, and develop a sense of group identity (Forsyth, 1999).

Organizational Context of Team-Building

Team-building requires examining the organizational context of the team
(Hayes, 1997). Many organizations trying to implement teamwork fail to

appreciate how their current practices and culture limit the ability of teams to operate. Effective teamwork requires a supportive organizational environment. It may not be useful to focus on the internal problems a team is having if the source of the problems lies in the surrounding organization. For example, it is difficult to promote cooperation among team members when the performance evaluation and reward system focuses on individual accomplishment and ignores team accomplishment.

The context in which the team operates has a greater impact on performance than the internal competencies of team members (Mohrman et al., 1995). Therefore, team development must focus on building the relationship between the team and its organizational context. Too often, a team-building program focuses on internal development when the key to performance problems is an external one.

A central organizational context issue for a team is the organization's performance evaluation and reward system. Performance evaluation systems have the potential to provide a team with feedback that can be used to analyze the team's operations and develop improvements. Reward systems provide motivation for the team members to work together and strive for efficacy. Although a performance evaluation and reward system typically is not considered part of a team-building program, it is an effective way for an organization to encourage and motivate a team to improve the way it operates. (Chapter 18 discusses evaluating and rewarding teams in more detail.)

Evaluating Team-Building Programs

Although many organizational leaders say that teams are important and team-building is an important activity, their actions do not necessarily support their words (Dyer, 1995). Most companies that use teams do little in the way of team development. They fail to include teamwork in their philosophies and reward systems. Top management focuses on financial issues, often ignoring the value of teamwork as a means of improving performance. The result is mixed signals about teamwork and the value of team-building.

Organizations overlook the importance of team-building activities for several reasons. They are limited by lack of expertise in team-building and the availability of competent professionals who can run team-building programs. Managers often do not understand the benefits of team-building, so they do not reward teams that spend time doing it. Team members are skeptical about the value of team-building, so they are reluctant to spend their time doing it.

Part of the problem in gaining organizational support for team-building programs is their reputation. Organizational programs are subject to fads, and team-building programs have suffered from it. During the 1960s, encounter groups encouraged people to share their true feelings with one another. Establishing good social relations among team members requires open communication, but the approach can be carried too far. Overemphasis on self-disclosure may create problems for group members in work settings. For example, perhaps a member's boss should not know one's inner feelings about him or her. Team-building through wilderness experiences—another fad—is enjoyable, but it can be difficult to translate these experiences into solving work problems. Rumors about poorly run team-building programs of the past limit support for team-building.

Most evaluation studies of team-building examine the internal operations of teams, measuring factors such as communication, cohesion, and satisfaction (Sundstrom, DeMeuse, & Futrell, 1990). Although these factors improve with team-building, the impact on team performance is often not measured. The relatively few studies that have measured performance usually find that team-building does not improve performance. Team-building interventions that focus on team control over work are more effective at improving performance than morale-boosting activities (Cotton, 1993).

For team-building programs, success depends on several factors (Dyer, 1995). Top management must provide support for teamwork and team-building programs. Organizational reward systems should support the use of team-building to encourage team members to take the programs seriously. Time must be made available for teams to engage in team-building activities. Finally, team-building programs are improved when they are linked with actions to improve the external relations of the teams.

2. Does Your Team Need Team-Building?

To determine whether team-building is needed, criteria for effective teams and an understanding of the problems that may interfere with team performance are required. Table 17.1 presents a set of criteria that may be used to evaluate the effectiveness of teams.

Effective teams have clear goals and values that are understood and accepted by all team members. Goal setting helps provide the objectives and task assignments. Team members must understand their assignments and how their roles fit into the team's activities. The team climate provides trust and support among team members so that they are willing to share ideas and feelings. All team members participate in the team's communication processes, and the team strives to make a majority of decisions through consensus. Once decisions have been made, team members accept them and commit to implementing the decisions. Leaders are supportive of team members and help facilitate team processes. Differences of opinion are recognized and handled

TABLE 17.1

Criteria for Effective Teams

- Clear goals and values.
- People understand their roles and assignments.
- Climate of trust and support.
- Open communication.
- Full participation in decisions.
- Commitment to implement decisions.
- Supportive leaders.
- Constructive handling of differences.
- Structure consistent with goals, task, and people.

SOURCE: From Dyer, W. *Team Building: Current Issues and New Alternatives* (1995). Reprinted by permission of the author.

rather than ignored. Finally, team structures and procedures are consistent with how the team operates and what it wants to accomplish.

Table 17.2 presents a set of symptoms of team problems that indicate when team-building is needed. Many of these symptoms identify the effects rather than the causes of problems (Dyer, 1995). In many cases, the two

TABLE 17.2

Symptoms of Ineffective Teams

- Loss of production.
- Increase of grievances or complaints.
- Evidence of hostility or conflicts among members.
- Confusion about assignments and relationships.
- Decisions misunderstood or not enacted.
- Apathy and general lack of interest.
- Lack of initiative, innovation, or effective problem solving.
- Ineffective meetings.
- High dependency on the leader.

SOURCE: From Dyer, W. *Team Building: Current Issues and New Alternatives* (1995). Reprinted by permission of the author.

causal factors are conflicts between team members and leaders and difficulties among team members. Conflicts with team leaders often lead to overconformity, resistance to leaders, an authoritarian leadership style, and lack of trust. Problems among team members often lead to bickering, lack of trust, personality conflicts, disagreements (with limited attempts to resolve them), cliques or subgroups, and missed deadlines.

Team members may blame individuals for team problems rather than recognizing that the team process is responsible. Team conflicts and confusion are good examples of problems blamed on individuals but that are really the responsibility of the team. Such problems can be addressed through team-building.

For example, how does one deal with two team members who are constantly arguing with each other? If the conflict seems to be due to a "personality clash," there is no solution except to get rid of one or both of the team members. An individual's personality cannot be simply rearranged to prevent future clashes. A more useful approach is to view conflict as a violation of expectations about what is to be done and how and when. Expectations focus on behavior, and behavior is open to change. The team can discuss its expectations about performance and can negotiate an agreement that will allow conflicting team members to work together.

Similarly, confusion among team members typically arises from unclear assignments and relationships. Solving this problem focuses not on individual team members, but on the team processes. Teams must better clarify the roles of all team members and not blame individuals for failing to perform unarticulated assignments.

3. Types of Team-Building Programs

Categorizing the programs typically used in team-building is difficult because there is no agreed-upon set of techniques. The following subsections describe five different types of team-building programs: goal setting, role definition, interpersonal process skills, cohesion-building, and problem solving.

Goal Setting

The goal-setting process is designed to clarify the purpose of the team. This approach involves clarifying the team's goals and developing more specific objectives. This typically is done through consensus building so as to create agreement about and commitment to the team's goals. Objectives are developed to further define the team's tasks and establish action plans that include team member assignments. The final step in a goal-setting program is to develop an evaluation and feedback system so the team can monitor the attainment of its goals (Locke & Latham, 1990).

A goal-setting program may be narrowly focused on the team's immediate performance or broadly focused on the values and mission of the team and organization. The broader view is designed to develop a common vision for the team by exploring the team's underlying values and purpose. The broader approach is useful for teams that will exist for a long time and for teams whose members come from diverse backgrounds (Hayes, 1997).

Role Definition

The role-definition approach focuses on clarifying individual roles, group norms, and shared responsibilities of team members. Conflict between roles and ambiguity about roles may create stress and disrupt performance. Team members need to be clear on both their own roles and the roles others perform. In clarifying duties and task relationships among team members, coordination is improved and the team is better prepared to perform its task.

Several approaches may be used to enable a team to define its roles (Hayes, 1997). The negotiation approach asks team members to analyze their work situations and identify what other people could do to improve their effectiveness. This includes behaviors that they would like to see more and less of. The team members then negotiate changes in one another's behaviors to get what they want from the other team members.

An alternative approach asks team members to interact and group process observers to analyze the roles they perform. These observations are used to evaluate the team's performance. The typical result of analysis shows the team is underusing certain behaviors and relying on a limited range of behaviors. This information may be used to help improve team interactions.

The value of role-definition approaches is that they allow team members to see themselves from the outside, through the eyes of an observer or other members. This allows members to develop a new perspective for understanding how they operate and teaches them how to adjust their interaction styles in ways that improve team operations.

Interpersonal Process Skills

Team members need to learn how to coordinate their efforts with other members and work together as a team. Decision making, problem solving, and negotiating are some of the team process skills members can learn. Given that a team's task problems are related to lack of these teamwork skills, teaching members interpersonal process skills is one approach to team-building.

Teaching group process skills is more than just lecturing. In team-building, a team typically is given simulated activities or exercises to perform to practice these skills and analyze the results (Scholtes, 1988). For example, rather than waiting until it has to make an important project decision, the team may practice decision-making techniques in a desert survival exercise.

Process consultants are often used to facilitate these group exercises, observe how the team operates, and comment on the group process (Forsyth, 1999). Feedback from these outside observers is viewed as key in the learning experience. This approach to learning has traditionally been viewed as the best way of improving group skills.

Cohesion-Building

The purpose of cohesion-building is to foster team spirit and build interpersonal connections among team members. When successful, it strengthens the team's morale, increases trust and cooperation, and develops group identity. Cohesion-building is accomplished through techniques that create unity and a sense of belonging, a climate of mutual understanding, and pride in the team (Hayes, 1997). The goal is to increase the sense of being a part of the team. Once team members have firmly established this relationship, they are more committed to the team goals and more supportive of the actions of other members.

Creating unity helps develop a sense of cooperation and belonging. One technique is to identify team boundaries so that team members have a greater sense of being part of something unique. Once they begin to see themselves this way, they will notice more similarities among fellow team members and more differences from outside groups. These psychological distinctions encourage team identification and commitment.

Building pride in the team builds relations among team members. Because pride is enhanced by professionalism, skills training can improve a member's image of the team. Various techniques, such as celebrating team successes, are concrete demonstrations of their accomplishments and build pride in the team.

One popular cohesion-building activity is an outdoor experience program. In this program, the team leaves its work environment and meets in an outdoor setting. Team members are presented with a series of challenges they must deal with as a team, such as crossing a river using ropes or climbing a mountain. In working together to meet these challenges, the team develops a sense of cohesion and accomplishment.

Problem Solving

Team-building is designed to improve team operations. Rather than starting with an approach to team-building, the team may start with an analysis of its problems. The problem-solving approach to team-building starts with problem identification and analysis. The problems may come from performance data, objective outside sources, or internal team communication. Information about team problems is gathered through surveys, interviews, or group discussions. This information is organized and shared with the team. Early on, a diagnostic session clarifies the problems and identifies the team's strengths and weaknesses.

The diagnosis stage ends with a discussion of how the team should proceed to take action on solving its problems. A standard problem-solving approach is used to generate alternatives and develop solutions. An action plan is developed for implementing the proposed changes in how the team operates.

This sounds like a fairly straightforward approach to solving a team's problems, but it may be difficult for a team to conduct. By the time the team recognizes it has a problem, the underlying conflict may make resolution difficult. Outside consultants are often needed to help the team analyze itself, develop alternatives, and negotiate acceptable solutions.

4. Team Training

Teams do not always live up to expectations because of lack of training in how to operate (Marks, Sabella, Burke, & Zaccaro, 2002). Teamwork training often focuses only on the development of interpersonal skills. However, team members need to learn more than just a set of generic teamwork skills. To be effective, team members must understand their roles, coordinate their actions with others, and understand how their actions interact with others (Goldstein & Ford, 2002). Training is required to develop task-related skills and knowledge, teamwork skills, and knowledge of the skills and roles of the other team members. Finally, teams need training on process improvement skills so they can learn how to improve performance.

A training program starts with a needs assessment to determine a team's training needs and objectives (Arthur, Bennett, Edens, & Bell, 2003). Conducting a needs analysis to determine specifically what training the team needs makes the training program more focused. The process starts with an assessment that analyzes the goals of the team, requirements of the members, the tasks they must perform, and the types of coordination needed. The information is used to create training objectives and design training programs. After training is conducted, evaluations are used to determine whether training objectives have been met.

The effectiveness of training programs depends on the method of training, the type of skill being learned, and the training environment (Arthur et al.,

2003). Teamwork training is more effective when it is focused on identified training competencies or skills, the team is trained together, they have an opportunity to practice their new skills, and they are given feedback about their performance. This ensures that the training will be transferred from the training session to the work environment. Two important issues related to effective teamwork training programs are training the team together and planning for the transfer of training.

Training Together

Training of team members should be done with the team as a whole; this develops the team's mental model or transactive memory. A team's mental model is a common understanding by team members of how the team operates (Klimoski & Mohammed, 1997). Effective teams have a shared understanding of the team's goals, norms, and resources, including understanding of the roles, knowledge, and skills of each team member. This shared understanding is essential for coordination among team members, especially for tactical teams (teams that carry out a procedure, such as a surgery or aircraft team).

Unfortunately, teamwork training focuses on individuals rather than groups. People receive training in teamwork skills and then are expected to apply the new skills when they join a team.

Transfer of Training

Transfer of training refers to the extent that the new skills learned in training are applied, generalized, and maintained in the work environment (Salas & Cannon-Bowers, 2001). Factors that affect transfer of training include the environment where the training occurs, the time lag between the training and opportunities to apply the skills, situational cues in the job environment that prompt applying the new skills, and social and supervisor support for their application. The transfer of training problem is important; research suggests that only 10% of training actually transfers to the work environment (Smith-Jentsch, Salas, & Brannick, 2001).

Several training strategies may be used to encourage transfer of training, such as practice and feedback in simulated environments, encouraging supervisor support of the application of new skills, and creating a supportive team climate. Transfer of training is easier when the application environment is similar to the training experience. The team leader may provide opportunities to perform the newly learned skill, model the skill, and reinforce attempts at performance.

The team climate refers to the degree to which the team expects and supports attempts to use the newly learned skills. Training people in the teams in which they will be working creates a supportive team climate for new skills performance. The supportiveness of the work environment has a substantial effect on whether the newly learned skills are applied (Edmondson, Bohmer, & Pisano, 2001).

5. Types of Training

Team members may only receive training in generic teamwork skills. This training is similar to other types of training provided by organizations. Several types of training have been developed specifically for teams, three of which are presented here: team resource management training, cross-training and interpositional training, and action learning.

Team Resource Management

Team Resource Management (TRM) (also called Crew Resource Management) is a training program to develop a defined set of teamwork competencies so a team can operate without error under stressful circumstances (Goldstein & Ford, 2002). This approach was developed for the aviation industry and has application to other action teams that perform in stressful situations, such as military and surgical teams.

TRM begins with a methodology for analyzing the team to identify its mission requirements and coordination demands. The information is used to develop a set of team competencies for training. A training method is then selected and exercises are developed for team members to practice their new skills. The exercises provide an opportunity to evaluate the effectiveness of the training program and to provide constructive feedback to the participants. The teamwork competencies and skills needed by teams that are often the focus of training are presented in Table 17.3.

TRM training combines classroom training with practice in simulators with feedback (Salas, Bowers, & Edens, 2001). It focuses on developing identified teamwork competencies necessary to the team, and has developed a set of tools and methods for identifying and teaching needed competencies. TRM training strategies use evaluations to encourage development of effective training programs. This training has been shown to be successful in reducing errors and accidents, improving teamwork, and increasing efficiency. The approach to training is effective because it focuses on specific

TABLE 17.3

Teamwork Competencies and Skills

Competencies and Skills	Definition
Adaptability	The ability to use information from the environment to change how the team operates.
Shared Situational Awareness	The development of an understanding of the team's internal and external environment and dynamics.
Performance Monitoring and Feedback	The ability to accurately monitor performance, provide constructive feedback, and use this information to improve operations.
Leadership and Team Management	The ability to direct and coordinate the activities of the team.
Interpersonal Relations	The ability to facilitate interactions to resolve disputes and motivate performance.
Coordination	The ability to organize the team's resources and actions so that performance is effective.
Communication	The ability to effectively exchange information with other team members.
Decision Making	The ability to gather and analyze information and use this information to solve problems.

SOURCE: Adapted from Cannon-Bowers, J., Tannenbaum, S., Salas, E., & Volpe, C., Defining competencies and establishing team training requirements. In R. Guzzo & E. Salas (Eds.), *Team effectiveness and decision making in organizations.* Copyright © 1995, San Francisco: Jossey-Bass.

teamwork competencies and provides trainees with opportunities to practice their skills and receive feedback.

Assertiveness training is an example of the type of training that is part of TRM. Problems with teamwork are a chief cause of major air carrier accidents (Jentsch & Smith-Jentsch, 2001). Accidents occur when crew members are unwilling to communicate problems to their superiors. This lack of assertiveness is a major problem for airline crews, medical teams, and police and firefighting teams, which is why assertiveness training is a standard element of TRM training.

Assertiveness is both a skill and an attitude. Consequently, training programs focus on both developing assertive communication skills and changing attitudes about when assertiveness is appropriate. Assertiveness is a situation-specific behavior. People's willingness to be assertive depends on the situation they are in and their relationship with the other people involved. This is why it is important to conduct assertiveness training in environments that match the occasion when the assertive behavior should be applied. Acting assertively is highly dependent on the team culture established by the team leader. Training programs should target team leaders in how to appropriately acknowledge assertive communications from subordinates. Behavioral skills training includes role-play practice in situations similar to those of the work environment. Active practice and feedback about performance is essential to learning new communication skills.

Cross-Training and Interpositional Training

Cross-training is used to increase the flexibility of team members (Goldstein & Ford, 2002). In cross-training, team members are trained in the technical skills of two or more jobs, allowing the team to assign members to the tasks that need to be performed in the present and supporting self-management of teams. In some cases, a pay-for-skills program is used to reward team members for learning new skills. A typical example of cross-training is a manufacturing team in which members learn multiple roles. This allows the team to flexibly respond to changes in the work environment and personnel changes. Cross-training programs often use on-the-job training, with experienced team members training other members.

The goal of interpositional training is to allow team members to better understand the working knowledge and roles of other team members and the interconnections among the actions of team members (Goldstein & Ford, 2002). This type of training leads to shared mental models of how the team operates and the coordination needs of the team; it should help facilitate interactions among team members and improve coordination. An example of interpositional training is teaching a flight crew to better understand one another's roles and abilities but not to replace each other's positions.

Interpositional training is designed to develop shared knowledge structures among team members (Marks et al., 2002). This is especially significant for action teams with highly interdependent work arrangements. Action team members have specialized skill sets (such as the members of a surgical team), rely on coordination to complete their tasks, and operate in challenging environments. Because of their interdependence, effective performance requires coordination and successful interaction among team members.

Three types of interpositional training vary in depth of knowledge and method. *Positional clarification* simply provides team members with information about the tasks and roles of other team members. *Positional modeling* adds observation of team members' duties to this information. *Positional rotation* has a team member perform another member's job to gain personal experience with the other's duties. The most appropriate form of interpositional training depends on the level of interdependence in the team roles. Positional clarification is useful when there is some interdependence; positional modeling may be preferable when teams are highly interdependent.

Interpositional training improves communication among team members and team performance. It has this effect because the training develops shared mental models among team members. Shared mental models improve team coordination, which improves performance, especially in interdependent tasks. Interpositional training is not designed to teach members to perform each other's tasks, but to raise awareness of others' roles. In addition to improved coordination, interpositional training increases "back-up" or assistance to team members in performing tasks. Although interpositional training is especially important for action teams, it is beneficial for other teams. Service teams, for example, benefit from having a greater understanding of one another's roles to coordinate service to customers.

Action Learning

The action learning approach is based on the belief that most learning occurs when people are directly dealing with real-life issues (Goldstein & Ford, 2002). The focus of action learning is to develop teams that can analyze and solve important, real-life problems in their organizations. Action learning combines training with developing innovative solutions to existing organizational problems. In the process of solving the problem and reflecting on the team's problem-solving strategies, team members learn how to create and operate as effective problem-solving teams. The components of an action learning program are presented in Table 17.4.

Action learning is self-managed learning (Marquardt, 2002). The team gains skills and knowledge, then shares this knowledge with the organization. Team learning starts with generating knowledge by analyzing a complex issue or problem, taking action to solve it, and then evaluating the results. The team will be able to learn and perform better in the future because it has learned how to experiment with new approaches to problems and communicate its knowledge. It is a strategy that develops a team and promotes learning throughout an organization. Through thoughtful, innovative action, teams learn how to effectively use their resources.

The action learning approach simultaneously solves a complex organization issue while developing teamwork skills in people and organizations

TABLE 17.4

Components of an Action Learning Program

1. *Problem or challenge*	Learning is built around a problem or issue, a task that needs solving that is important to the team and organization. Teams learn best from taking action and reflecting on the results. The problem must be real and valuable and be capable of being affected by the team's efforts. Examples of problems include reducing turnover, improving quality, reorganizing a department, or improving an organizational process.
2. *Learning team*	Teams should be relatively small; from 4 to 8 members are optimal. It is important to have a diverse perspective, so cross-functional or cross-organizational teams are best. Teams must have the ability to implement actions and affect the organizational issue they are working on.
3. *Learning coach*	The learning coach facilitates the group process, encourages reflection, promotes communication, and facilitates problem solving. The coach asks, What are you learning, how can you solve this problem? The coach helps the team solve its problem but also promotes reflection on the group process and learning teamwork skills.
4. *Insightful questioning and reflective listening*	The coach encourages questioning and discussing the problem rather than jumping to a solution. This emphasis on analysis of the problem encourages reflection, creativity, and better problem-solving skills.
5. *Taking action*	There is no learning without action. The implementation and evaluation/feedback process is the key to learning. It is the reflecting on action where learning occurs.
6. *Commitment to learning*	Solving organizational problems is good, but the more important organizational benefit is the understanding gained of how to solve problems. Action learning promotes organizational learning and development. It is an approach for dealing with organizational problems, and it helps develop people with the skills to solve current and future problems. It promotes learning how to learn.

SOURCE: Adapted from Marquardt, M., *Building the Learning Organization*. Copyright © 2002, Palo Alto, CA: Davies-Black.

(Marquardt, 2004). It is an approach to continual learning that encourages experimentation, allows mistakes, offers support, and promotes feedback. It is a shift from a culture of training (where someone else determines what you need to know) to a culture of learning (where you are responsible for your development).

Summary

Team-building is a type of organizational development that focuses on improving the operation of teams. To be effective, it should be an ongoing team activity. Team-building examines both the internal processes and organizational contexts of teams. Although team-building programs have been shown to be effective, organizations and teams are often reluctant to conduct them.

Deciding whether a team needs team-building requires criteria for effective and ineffective teams. Effective teams have clear goals and tasks, open communication climates, supportive leaders, and procedures for managing tasks and problems. Ineffective teams have unresolved conflicts and hostilities, confusion about goals and tasks, low levels of motivation, and high dependence on their leaders.

There are many different types of team-building activities. Goal setting is used to clarify a team's goals and objectives. Role definition clarifies individual roles and aids in establishing group norms. Interpersonal process skills use training activities to teach members to work together as a team. Cohesion-building tries to create a team identity and improve social relations among team members. Problem solving identifies the team's main problems and works with the team to develop and implement solutions.

Training programs may be used to develop task and teamwork skills, increase awareness of the roles in the team, and provide skills to improve performance. Training is more effective when based on a needs assessment and improved through evaluations. Team members should be trained together and the training integrated into the work environment.

There are three main types of teamwork training. Team Resource Management (TRM) uses assessment to identify needed teamwork competencies and combines training with practice and feedback to ensure its effectiveness. Cross-training and interpositional training improve team flexibility and coordination by enabling team members to understand the roles and interconnections in the team. Action learning uses participation in facilitated problem-solving teams to develop teamwork skills and solve important organizational problems.

TEAM LEADER'S CHALLENGE—17

You are the team leader at a large engineering design firm. The general manager of the firm is extremely frustrated with his attempts to develop and use project teams. He has tried personality testing to improve communication skills; wilderness experiences to improve problem-solving skills; and Japanese culture lessons to improve consensus decision making. None of these approaches has worked, and he now wonders whether engineers can be trained to work in teams.

The most recent team-building program is using "professional" team facilitators who are graduate students in counseling from a nearby university. Their use is getting a mixed reaction from team members. (Maybe learning to better express one's feelings wasn't what these teams needed.) The facilitators are often effective in improving the performance of the teams, but they are not encouraging them to learn how to operate independently. The director is looking for advice from his project team leaders about the use of teams and the future of team-building in the firm.

What has gone wrong with past team-building efforts?

What advice would you give the director about team-building in the organization?

How can action learning be used to help promote teamwork in this organization?

ACTIVITY: TEAM-BUILDING

Objective. Team-building programs typically promote team development and learning by having the team perform some activity while observers record the group process. After the activity has been completed, a facilitator discusses the group process observations with the team and uses this information to help the team improve how it operates.

Activity. Form teams of five to seven people and distribute a copy of the "University Discipline Board Decision" (Activity Worksheet 17.1) to each team. Select one or two observers and give them one of the group process observation forms found at the end of the chapters in this book. For example, Chapter 6—Communication network, Chapter 8—Power styles, Chapter 9—Consensus rules, and Chapter 13—Diversity. After the team has discussed the problem and reached a consensus decision, ask the observers to provide feedback to the team.

ACTIVITY WORKSHEET 17.1

University Discipline Board Decision

You are members of a University Discipline Board. You have been asked to review the following case and determine whether the accused student violated the university's Sexual Harassment Code or any other relevant University Code of Conduct. If you find the student guilty, you must decide on the appropriate punishment. All decisions made by the University Discipline Board must be unanimous.

Case:

John Smith is a sophomore at the university. He is taking an English class in creative writing. He decided to write a series of stories about a serial rapist who violently preys on college women at an anonymous university. In his stories, the names of the women victims are students in his writing class. He decided to "publish" his stories by putting them on the Internet in his electronic homepage (which is based on the university's computer). Several of the women students in the class read the stories, were upset by them, and complained to university administrators.

Decision:

1. Is John Smith guilty of violating the university's Sexual Harassment Code or any other relevant University Code of Conduct?

   _____ Yes   _____ No

2. If yes, what disciplinary action should the university take?

Analysis. Use the analysis sections related to the observation sheets to discuss the group process. This activity may create gender differences, so examine how the gender mix of the teams affected their ability to reach an agreement.

Discussion. How can using team-building activities such as this help to improve a team's group process? Examine the activity on Norms for Virtual Teams at the end of Chapter 15. What norms should a university develop about the use of information technology?

18

Evaluating and
Rewarding Teams

An important way to motivate teams is through performance evaluation and reward programs. Performance evaluations communicate to the team how well it is performing. This information may be used to provide direction, improve and motivate performance, and inform the organization's reward system. Team members should participate in the evaluation process given that they are most aware of the contributions of an individual team member.

The shift to teamwork often requires organizations to change the way they reward people. The individual reward programs used in traditional organizations may not reward commitment and participation in teams. Organizations may use combinations of individual, team, and organizational rewards to motivate teams. The best reward program depends on the nature of the task and the type of team.

Learning Objectives

1. How does the performance evaluation system affect teamwork?

2. What are the characteristics of a good team performance evaluation system?

3. What are the advantages of using multi-rater performance evaluations?

4. How do biases (such as the team halo effect) influence the evaluation process?

5. Why do organizations need to change their reward systems to support teamwork?

6. What are the advantages and disadvantages of individual versus team rewards?

7. Understand the main types of individual, team, and organizational rewards that are useful for teams.

8. How does the best type of reward program relate to the type of team?

1. Team Performance Evaluations

Team performance evaluations provide feedback to the team to improve the way it operates and may be linked to rewards to motivate team members. Organizations may evaluate individual team members, the operation of the entire team, or combinations of individual and team. Performance evaluation measures need to be specific and clear, identified in advance, and related to behaviors under members' control. The evaluation process should include participation from both supervisors and team members. Although biases are created by participating in the evaluation process, multi-rater evaluations increase the accuracy and acceptance of evaluations.

Developing a team-oriented performance evaluation system and using the information to provide feedback is an important way to improve team operation (Mohrman, Cohen, & Mohrman, 1995). Teams use feedback from performance evaluations to identify and correct problems in operations. In addition, performance evaluations are used to measure the team's success and provide input to the organization's reward system, which motivates the team to better performance.

Performance evaluations are valuable for providing feedback to employees, motivating them, and supporting training and development. Unfortunately, the evaluation process often creates conflict and leads to dissatisfaction rather than motivation and development. Because of this potential for conflict, managers and employees try to avoid conducting performance evaluations. Managers may feel uncomfortable about giving feedback to subordinates, and most employees believe their performance is above average, so they resent receiving constructive criticism (Lawler, 2000).

Types of Evaluations

Three main approaches to team performance evaluation are: traditional individual evaluations, team member evaluations, and evaluations of the team (Lawler, 2000). In traditional evaluations, the supervisor appraises an individual employee; this is tied to the organization's compensation system. Performance evaluations are typically done this way in most organizations. In the second approach, team members conduct the performance evaluation instead of a supervisor. When the work of a team is highly interdependent, it may be impossible to evaluate employees individually. In such a case, the supervisor, not the individual team members, evaluates the operation of the entire team.

There are several factors that affect which type of performance evaluation is appropriate (Lawler, 2000). From a teamwork perspective, the most important factor is work design. Individual evaluations are most appropriate for individual work assignments. When work is highly interdependent or conducted primarily in a team, individual evaluation approaches are not appropriate and may discourage cooperation in the team. However, the type of evaluation used often depends on how the organization operates. Most traditional organizations rely on individual performance evaluations linked to compensation programs, even when employees primarily work in teams.

For some teams, such as sports and action teams, evaluations may include both individual and team performance measures. Individual performance measures relate to the organization's reward system, reduce social loafing and other motivation problems, and provide information useful for identifying the assistance needed for specific individuals. For other teams, such as production and project teams, evaluations may measure only the team's performance because of the difficulty in accurately measuring individual as distinct from team performance.

Types of Measures

The key to developing a good measurement system is making certain that it captures both team and organizational goals. A lack of clear team goals and accountability is one of the main reasons for the failure of work teams (Jones & Moffett, 1999). Team performance measurements should relate to contributions to the organization. It is necessary to ensure the measurements relate to factors that the team has the ability to influence (Zigon, 1997). The measures should focus on the results of the team's performance, not on the internal activities of the team, because the team should be free to accomplish its goals the way it wants to.

Performance evaluations are improved when specific, quantifiable goals are set for the team, and these are identified in advance (Lawler, 2000). People need to know, in advance, how their performance will be measured and the acceptable level of performance to make the performance evaluation process fair and motivate employee performance. Without goals and accurate measurement criteria, team members do not know how to act to receive positive evaluations. It is useful to include team members in the design and development of the performance evaluation system. This increases understanding of the evaluation measures and acceptance of the evaluation process.

Many performance evaluations systems fail because the measures are vague or poorly defined (Lawler, 2000). Rather than evaluating traits such as reliability, cooperativeness, or leadership, it is better to use behavioral or outcome-based measures that can be quantified and clearly defined. For example, rather than cooperativeness, ask about the frequency of participation in team meetings, providing assistance to team members, and so forth. Vague measures are more vulnerable to stereotypes and other perceptual biases.

When developing a team performance evaluation system, one basic alternative is using behavioral measures versus the results of performance (Rynes, Gerhart, & Parks, 2005). Behavioral performance measures have advantages: They can be used for any type of job; the rater can deal with factors outside of the employee's control that impact results; and they can encourage positive employee behaviors (like cooperation) that are not directly related to the job. Although results-based evaluations are more objective, high-quality objective measures are unavailable for many teams. Consequently, most performance evaluations include a mixture of these two types of measures (Gross, 1995).

Typically, a team performance measurement system contains 5 to 10 different measures (Jones & Moffett, 1999). Because the purpose of measurement is to provide feedback and improve performance, it is preferable to have a simpler system team members can relate to, rather than a sophisticated system that is hard to interpret. Using behavioral scales encourages the rater to focus on the target's behavior rather than personality. Including team members in the development of behavioral scales is a good way to improve the relevance and credibility of the evaluation system. Table 18.1 presents a simple behavioral scale developed by students for evaluating contributions to team projects (Levi & Cadiz, 1998).

Participation in the Evaluation Process

Traditionally, the employee's supervisor conducts the performance evaluation; however, teamwork evaluations should be performed by more than just the supervisor (Lawler, 2000). It is difficult for a supervisor to conduct a performance evaluation on someone who works on a team. Teamwork often requires a shift to the use of multi-rater evaluation approaches, such as

TABLE 18.1

Behavioral Rating Scale to Evaluate Student Team Projects

Use the following rating scale to evaluate your team members' behavior on the class project:

1 - never   2 - sometimes   3 - usually   4 - always

Did the team member you are rating:

    A. Make commitments to do tasks

    B. Do his or her fair share of the work

    C. Produce work with acceptable quality

    D. Actively participate in team discussions and decision making

Rate each team member:

Team member	A	B	C	D	Total
1. _____	___ +	___ +	___ +	___ =	___
2. _____	___ +	___ +	___ +	___ =	___
3. _____	___ +	___ +	___ +	___ =	___
4. _____	___ +	___ +	___ +	___ =	___

360-degree feedback, that include input from team members, customers, and supervisors. Those most qualified to evaluate a team member's performance are his fellow team members. The team's supervisor and customers can evaluate the overall performance of the team, but only the team members can accurately evaluate an individual's role in the team (Gross, 1995).

Multi-rater performance evaluations are more reliable and valid than supervisor-only evaluations (Rynes et al., 2005). Also, people are more likely to find feedback from multiple sources to be credible and fair. Sometimes performance evaluations from the supervisor are used for pay decisions, while evaluations from team members are used for feedback and development purposes (Lawler, 2000). This makes it easier for the team member raters to be honest because their feedback does not affect pay decisions for the recipient.

A multi-rater approach may be used for student team projects. Students are more accurate evaluators of their performance on team projects than professors (Levi & Cadiz, 1998). The professor only sees the final product, while the students observe the performance of each team member. However, professors are reluctant to use student evaluations because of a belief that students will not accurately evaluate one another. The use of student evaluations of team projects reduces social loafing by giving students the ability to influence their teammates' behavior and increases the perceived fairness of the evaluation system.

Problems and Biases With Team Evaluations

Team performance evaluations should not encourage competition among team members. Many organizations use evaluation systems that require a fixed distribution of good and poor performance (for example, 20% higher performers, 60% moderate, and 20% poor performers). These practices are inappropriate for teams (Lawler, 2000). Ranking or classifying employees against one another encourages competition among employees and reduces teamwork. If the team is performing well and everyone is doing their fair share, ranking team members against one another is inappropriate and can damage team operations. A team that works well together should not have performance gaps among team members.

Including team members in the performance evaluation process raises questions about fairness and accuracy (Gross, 1995). Personal relationships and favoritism are a problem in evaluations, regardless of whether they come from a supervisor or a team member. Team members often do not have experience or training conducting performance evaluations, so they need help doing them. Although there are problems with team members' ratings, the main limit on use of team member evaluations is that managers are often reluctant to give up the power to conduct performance evaluations.

Although it is valuable to include team member evaluations in the performance evaluation system, some team members are uncomfortable evaluating coworkers' performances, especially when the evaluations are related to pay increases (Lawler, Mohrman, & Ledford, 1995). Team members feel more comfortable evaluating one another when they can use objective performance standards and the ratings remain confidential.

All evaluation systems that utilize people rather than purely objective systems may suffer from bias. Peers are no more likely to have biases than supervisors. Some biases are more likely to affect peer evaluations. For example, inflation bias refers to positive evaluations given when raters expect to have to present feedback to others. Inflation comes from both empathy and fear of creating conflict in the team (Antonioni, 1994). Reciprocity bias occurs when people feel obligated to give positive ratings when they have received positive ratings from others.

One type of bias is unique to teams. The team halo effect describes how team members view the success and failure of teams (Naquin & Tynan, 2003). When teams are successful, team members view success as caused by the team; when they are unsuccessful, they tend to blame individual members for the failure. This type of scapegoating is likely to affect team members' evaluations of one another. When the team is successful, there is a tendency to rate every team member as a good performer. When teams fail, members are more likely to give negative performance evaluations to selected members.

2. Reward Systems

An organization's reward system is an important way of encouraging a team to improve the way it operates. Team rewards have the potential to influence the motivation of individual team members, the level of coordination in the team, and the quality of the group process. More organizations are beginning to use some type of team rewards to encourage team effectiveness (DeMatteo, Eby, & Sundstrom, 1998).

The shift to teamwork creates problems with traditional compensation systems (Lawler, 2000). An organization needs to change its compensation practices to support teamwork, because pay is a communication device that tells employees what is important. However, compensation programs are typically conservative, so they are rarely used to lead organizational change. As organizations become more team-oriented, existing compensation practices either get in the way or are modified to support the use of teams (Lawler, 1999).

In traditional companies, pay systems focus on individuals, people are paid based on their position in the organization's hierarchy, and bonuses are reserved primarily for managers. These companies rely on increases in base pay (often by seniority) and promotions as a way of rewarding and motivating employees. But in team-based organizations, with flatter hierarchies and flexible job assignments, this traditional approach to compensation does not work effectively (Gross, 1995).

There are three approaches to rewarding performance: individual, team, and organizational. Individual reward systems are good for motivating high performers, but may discourage cooperation and teamwork. Team and organizational approaches are better at encouraging teamwork and are appropriate when the tasks are highly interdependent (Cohen & Bailey,

1997). There also are in-between options. For example, production workers and professionals are often evaluated and rewarded for individual performance. However, information about their participation in teams may be included in their individual performance evaluations.

Most U.S. employees prefer individual, rather than team-based, rewards, particularly employees who are highly achievement-motivated (Rynes et al., 2005). Team rewards tend to have less incentive effect than individual rewards; however, individual reward programs may not encourage sufficient cooperation among team members. When teams are rewarded collectively, the team has the right incentive to deal with members who are not doing their fair share. On the other hand, not offering rewards for individual performance can reduce the motivation of high performing team members (Lawler, 2000).

Although team rewards are valuable motivational tools, their use may create problems. An emphasis on team rewards may encourage social loafing, discourage the performance of good workers, and create inequity problems (DeMatteo et al., 1998). This is why organizations often use a hybrid approach that combines team and individual rewards. Teams that are rewarded for a mix of individual and team performance perform better than teams rewarded for solely individual or team performance (Fan & Gruenfeld, 1998). Because team rewards may foster competition among teams in organizations, organizational rewards are used to encourage cooperation among teams (Lawler, 2000).

Types of Teams

The type of team and its organizational level affects the reward program. Teams vary in whether they perform the basic work of the organization or supplemental work and whether they are temporary or permanent. The three basic types of teams are parallel, process, and project teams (see Table 18.2). Organizational level affects teams because of skills (common versus specialized) and power (ability to impact the organization). Skill-based pay makes sense for a factory team, but not for a professional team. Profit-sharing may be useful for a management team, but inappropriate for a service team.

Parallel teams are typically part-time, temporary teams that supplement the regular work system and perform problem-solving and improvement functions (Lawler, 2000). An example of a parallel team is a quality improvement team that meets to analyze and discuss issues of quality. Parallel teams often meet regularly, follow predesigned systems for problem solving and analysis, receive teamwork training, and make recommendations for approval by management. They provide ideas or recommendations for change, but do

TABLE 18.2

Types of Teams

Type of Team	Use	Characteristics
Parallel team	Problem solving Improvement	Temporary Partial commitment
Process or work team	Make a product Provide a service	Permanent Full commitment
Project team	Design a product Implement a program	Temporary Full commitment

not implement them. Such teams often have a limited lifespan and disband after working on a particular topic. Parallel teams are widely seen in industry, with more than 85% of large U.S. companies using them.

Parallel teams are a great way to encourage employee participation in the organization. However, they create an evaluation and reward problem for the team members (Gross, 1995). There is an inherent conflict between commitment to the team's task and commitment to the member's primary job functions. Members of parallel teams spend limited work time in the team and most of their time working for a supervisor who is not part of the team. Because of these dual loyalties, parallel teams often have problems motivating members. Including an employee's work on the parallel team in his or her performance evaluation helps deal with this dilemma.

The process or work team is a full-time, permanent team that is the major part of the members' work. Process or work teams are responsible for producing a product or providing a service (Lawler, 2000). Process teams range from factory workers assembling products to airline crews and surgical teams. Employees may have similar skills and training or be cross-functional. Process teams are used by 78% of large U.S. companies. They typically control how the task is performed, but vary in how much control they have over selection of tasks, personnel issues, and other work aspects.

Process teams have many benefits in performing complex tasks and providing social support to team members (Gross, 1995). However, they do create an evaluation and reward problem because it is often difficult to identify, evaluate, and reward individual behavior in the team, given that the task is interdependent. Measurement of performance of these teams is inherently at the team level. Organizations may use team rewards to motivate the entire team, but it is critical to recognize individual contributions as well; this often requires including team members in the performance evaluation process.

Project teams are typically composed of a diverse group of knowledge workers who come together to perform a project in a defined period of time (Lawler, 2000). An example of project teamwork is a new product development team. Project teams work on unique and uncertain tasks that require creativity and decision making for completion. Project teams are temporary structures whose membership may change during a project and disband when the project is completed.

Project teams require full-time commitment from team members, but are of limited duration. Although they complete a project, team members' quality of work is difficult to measure because its value may not be known until long after the work is completed. Project teams are often cross-functional, including team members with a variety of skills and expertise. They are highly effective ways to perform certain types of tasks. However, they create evaluation and reward problems given that there is often no supervisor capable of evaluating the specialized performance of the team member (Gross, 1995).

3. Rewarding Individual Team Members

In most organizations, rewards are based on job descriptions (Lawler, 1999). Specific jobs have salary ranges attached to them, and employees are paid on the basis of their jobs and periodic performance evaluations from their managers. One approach to making individual reward systems more responsive to teamwork is to change the performance evaluation system by including factors such as cooperation and team participation. Because managers often are unaware of the internal operations of a team, this requires input from team members in the evaluation process. Other approaches to supporting teamwork through individual rewards are changing how base pay is calculated and adopting a skill-based pay program.

Changing Base Pay

Team members' base pay should be fair and equitable. Because they share work, support one another, and respond flexibly to work situations, their base pay should be similar (Gross, 1995). In traditional organizations, each job has a different base pay. When people shift to teamwork, the distinctions among jobs can disappear. In team-oriented organizations, people do not have specific jobs, but roles and temporary task assignments (Lawler, 2000). Consequently, it makes more sense to pay people for what they are capable of doing (their skills and knowledge) than for the changing tasks they perform. This is why team-based organizations shift to broad pay bands rather than salaries for specific jobs. The policy reduces pay distinctions among team members, improving perceived equity and fairness.

The primary benefit of the team-oriented salary approach is that it gives the organization more flexibility for managing teams (Manufacturing Studies Board, 1986). Issues such as cross-training, job rotation, and teamwork do not cause compensation problems because salary is not based on the type of work performed on a given day. It gives the organization the option of deciding how much cross-training will be done, who and when to train, and where to assign employees without considering salary implications. Because individuals who prefer to perform limited job activities may continue to do so, while employees who seek job variety and want to learn new skills may be accommodated, this gives both the organization and individual flexibility.

The main problem associated with a team-oriented salary approach is that it eliminates one of the organization's means of rewarding individual performance. In the traditional compensation system, employees had an incentive to work hard and learn new skills in order to be promoted to a new job. The lack of an incentive for harder work is a significant problem, which often requires adopting a system of individual or team bonuses to motivate employees.

Skill-Based Pay

An alternative approach for supporting teamwork is to pay an employee for the skills or competencies he or she possesses (Lawler, 1999). Skill-based pay encourages employees to learn new skills, which makes them more flexible and increases their understanding of the work process (Gross, 1995). The primary benefit of this approach is to improve the flexibility of the team in performing its task. It is most commonly used when the team member skills are not too widely divergent and the team's actions are highly interdependent. For example, factory workers who are multiskilled can take over one another's jobs when bottlenecks occur or when team members are absent. Professional teams that provide an integrated service to customers, such as insurance or banking, use multiskilling to improve customer service.

The primary advantage of skill-based pay is organizational flexibility. However, several problems may occur with skill-based pay programs (Luthans and Fox, 1989). Training costs rise due to skill-based pay. There can be quality and productivity problems because employees are working on new jobs in order to gain new skills. Promotions may be limited after a few years because employees have learned all the skills offered by the company. Therefore, skill-based pay is often used to make the transition into work teams, and then abandoned once most of the workers have completed their skills training (Gross, 1995).

Knowledge-based pay or career ladders are a type of skill-based pay for professional workers (Lawler, 2000). Multiskilled professionals are a benefit to professional teams. Cross-training will not create the skills, but job rotation helps an organization nurture the development of professionals with skills and knowledge that cross traditional professional boundaries. Knowledge-based pay, which rewards employees for depth of knowledge in an area, is a useful organizational tool for developing and retaining technical experts.

4. Team and Organizational Reward Programs

Team rewards are based on a successful team performance. Organizations are adopting team rewards because using work teams makes it difficult to accurately evaluate individual performance as distinct from that of the team (DeMatteo et al., 1998). The effectiveness of team reward programs depends on the characteristics of the rewards, the organization, and the team (Gross, 1995; Lawler, 2000). Team rewards should be large enough to make a noticeable difference in the member's pay (often assumed to be about 10% of the salary). To encourage cooperation, rewards should be distributed equally among all team members. Team rewards are more effective when the organizational culture supports teamwork and when teams are integrated into the operation of the organization. Effective team rewards require clear team goals, measurable performance standards, and a task that requires integrated teamwork.

It can be difficult to develop appropriate team rewards for some types of teams. For example, project teams can be difficult to reward. An organization often rewards project team members when they successfully complete the project, which is a good team incentive. However, team members come and go during the course of a project, so it is difficult to determine who should share in the rewards. One approach is to use team recognition programs to reward project teams at important milestones in a project. An alternative is to rely on organizational rewards, such as profit-sharing, to reward successful teamwork.

Team Recognition Programs

Most companies do not have special reward programs for teams. When they do, the most common type is the use of recognition awards (Gross, 1995). Special awards and recognition programs may be used to reward successful team performance. Although recognition programs are a valuable way of acknowledging effort beyond expectations, these programs are useful to supplement, but not substitute for, team reward programs (Lawler, 2000).

Recognition awards are one-time events that acknowledge team and individual success (Gross, 1995). Recognition awards may be either cash or non-cash. Surveys show that about one-third of large U.S. companies use recognition awards to reward successful team performance. Non-cash awards should be appropriate for employees and given in a manner that publicly recognizes team accomplishments. Cash awards should be large enough (several hundred dollars or more) to have an impact.

When establishing a team recognition program, the decision must be made whether to give cash or non-cash rewards. There should be several levels or amounts of award to recognize different levels of contribution. To ensure fairness, there should be an organization-wide system for nominating and selecting teams for recognition. Awards should be given as timely as possible to a team success. The awards presentation should be a positive experience that publicizes the success of the team.

Organizational Rewards

A potential problem with team incentive programs is that they may inadvertently create competition among teams within an organization (Lawler, 2000). When an organization's work requires coordination among teams, the use of team rewards may reduce overall organizational performance. In addition, measuring the performance of individual teams may be difficult in some organizations. Profit sharing, gain sharing, and other organizational reward programs are a useful approach for dealing with these problems. Such organizational programs reward teams for helping an organization succeed or for achieving an organizational performance goal (such as quality, productivity, or customer service goals).

Profit-sharing programs are based on meeting profitability targets for the organization (Rynes et al., 2005). The large number of employees involved and factors outside of employees' control affecting profits may limit the motivational effects of these programs. Gain-sharing programs link pay to collective results at a level smaller than the entire organization. For example, they may be linked to the performance of a specific facility or production unit. This can make them more motivating, given that fewer employees are involved. They provide a way to give incentives to teams when jobs are interconnected and individual incentive programs cannot be used. Gain-sharing programs are powerful incentives that encourage teamwork throughout an organization (Rynes et al., 2005).

Organizational reward programs may be calculated and distributed to people in a variety of ways (Thomas & Olson, 1988). The success of the programs depends more on the organizational context and participants' beliefs than on the specific formulas used to calculate rewards. The biggest problem with organizational reward programs is the difficulty in establishing a connection between a team member's actions and the performance of the organization. This is especially true in larger organizations.

In U.S. culture, it is difficult to get away from individual reward systems. However, organizations may use multiple performance reward programs (DeMatteo et al., 1998). Most organizations use individual salary systems as their primary methods of compensation. Additional compensation may be based on skills, individual merit, team incentives, and organizational incentives. It is not really a matter of choosing which system is best, but selecting the right combination of approaches. The right combination is likely to depend on the goals and characteristics of the organization.

5. Relationship of Rewards to Types of Teams

The reward program an organization uses to support teamwork depends on the type of team being rewarded. Table 18.3 shows the relationship between types of teams and rewards.

Process or work teams benefit from the use of a variety of individual and team reward programs. Individual-based programs such as skill-based pay encourage the development of important team skills. Shifting to salary-based compensation, rather than paying people for the job they perform, encourages equality among team members. Team and organizational incentive programs motivate coordinated actions. Finally, individual incentives based on team member evaluations help motivate team members when the work is not highly interdependent.

Rewards for performance may not be needed for parallel teams; however, incentive programs may help motivate teams (Lawler, 2000). Organizational

TABLE 18.3

Relationship of Rewards to Types of Team

Type of Team	Reward Programs
Process or work team	Salary-based compensation
	Skill-based pay
	Individual, team, and organizational rewards
Parallel	Team recognition awards
	Organizational rewards
Project	Team recognition awards
	Team and organizational rewards

reward programs such as profit sharing and gain-sharing programs fit well with the use of parallel teams. However, there is a weak motivational link between the success of a parallel team and improved performance of the organization. Not only is it difficult for the team to see the connection between their work and organizational success, but success is shared with other organization members who are not on the team.

The alternative to organizational rewards is to give the team bonuses or recognition awards. One problem with using bonuses is that it is often difficult to estimate the value of the recommendations made by parallel teams, and the value is dependent on whether the ideas are accepted and implemented by management. Recognition awards are a popular alternative for rewarding parallel teams. The key to recognition programs is to give the team something it values when it has accomplished significant work.

Project teams may be the most difficult type of team to reward (Lawler, 2000). A common approach is to give the team a bonus or recognition award when its task has been completed. However, project teams may work for a long time on a project, and team membership may change during the project. In addition, member participation on the team may vary, so that giving each team member the same reward may be inappropriate. One alternative is to give team rewards or bonuses during project milestones. Team bonuses can be given equally, or relative to level of participation. Organizational reward programs such as profit sharing may be a useful way to reward project teams in smaller organizations.

Summary

Performance evaluations provide important information to help a team improve its operations and motivate performance by linking to the organization's reward system. Evaluations may focus on individual performance, team member evaluations of one another, or evaluations of the entire team. They may measure either the team members' behavior or the results of the team's performance. Evaluation measures should be developed through a participative process and be linked to both team and organizational goals. Participation of team members in the evaluation process is important because supervisors are often unaware of the internal operations of the team. Multi-rater approaches to evaluations that include team members increase the accuracy and perceived fairness of the evaluations.

There are problems and biases with any performance evaluation system. Using concrete behavioral measures reduces the impact of stereotypes and perceptual biases of raters. Confidentiality of evaluations reduces the reluctance of some team members to negatively evaluate others. Teams suffer from a halo effect wherein they give everyone high ratings when the team is successful and scapegoat selected members when the team fails.

Rewards may motivate individual and team performance. Often the traditional individual compensation system used by organizations gets in the way of teamwork. Individual reward systems are good at motivating individual performance but may reduce cooperation and commitment to the team. Team rewards encourage cooperation but may not encourage individual motivation. Organizational reward systems encourage cooperation among teams and commitment to the overall goals of the organization.

Team-oriented reward programs use a combination of individual, team, and organizational reward programs. The best approach to rewarding teams depends on the type of task and team. When tasks are interdependent, the organization should use team or organizational rewards. Process, parallel, and project teams differ in their amount of time and commitment to the team. This affects the best way to reward team participation.

Process teams require changes in how individuals are compensated to promote flexibility and skill development. Team reward programs are useful when the work of the team is so interrelated that individual performance evaluations and rewards are inappropriate. Team recognition awards are valuable ways of rewarding parallel and project teams, where team participation is not a full-time, permanent activity. Organizational reward programs, such as profit-sharing or gain-sharing, encourage teams in an organization to work together and reward teams that help the organization to succeed.

TEAM LEADER'S CHALLENGE—18

You are the leader of a quality improvement team that has worked to redesign your organization's work processes. The team is composed of members from various parts of the organization. Although this project is important for the organization, the team members only spend part of their time working on the project. The team has met regularly for the last six months. The commitment and motivation levels of the participants have been highly variable. Some team members complain that their managers do not consider the work of the team to be important.

The team presented its proposed changes last week to management, and the response was mixed. Clearly, it will be easier for some sectors of the organization to adopt and implement the recommendations than others. The benefits of the proposed changes would primarily help improve customer service, so it is difficult to measure the financial impact of the proposal. Your team has put in a lot of effort to develop this proposal, and you would like to see the members rewarded for their effort.

How should the team's performance be evaluated?

What is the best way to reward the team members?

How should the rewards be distributed?

ACTIVITY: EVALUATING AND REWARDING A PROJECT TEAM

Objective: Project teams can be one of the most difficult types of teams for which to develop effective evaluation and reward programs. It is important to balance evaluations of the results of the team's work with the contributions of individual team members. Finding the right balance and ways to measure these factors is key in a successful evaluation and reward program.

Activity: Use Activity Worksheet 18.1 to develop a performance evaluation and reward program for a project team. Select either a student project team or a professional new product design team as a specific example to use.

Analysis: Do the evaluation criteria provide sufficient direction to team members so that they know how to act in order to be successful? Will the reward program that you selected motivate team members to perform well? What problems are likely to be encountered if you use this program?

Discussion: What are the effects of using team member ratings to evaluate team performance? What is the right balance between individual and team rewards?

ACTIVITY WORKSHEET 18.1

Activity: Evaluating and Rewarding a Project Team

Team Goals: What are the main goals of the team?

_____

_____

Team Evaluation: What criteria should be used to evaluate overall team performance?

_____

_____

Member Evaluation: What criteria should be used to evaluate each team member's performance?

_____

_____

Reward Program: How should the team's work be rewarded?

What percentage of the reward should be based on the team's project versus the behavior of the individual team members?

_____ % team reward - based on supervisor's evaluation of the team project

_____ % individual reward – based on team members' evaluations of individual performance

References

Adler, N. (1986). *International dimensions of organizational behavior*. Boston: Kent.

Adler, P. (1991). Workers and flexible manufacturing systems: Three installations compared. *Journal of Organizational Behavior, 12,* 447–460.

Alberti, R., & Emmons, M. (1978). *Your perfect right*. San Luis Obispo, CA: Impact Publishers.

Allen, N., & Hecht, T. (2004). The "romance of teams": Toward an understanding of its psychological underpinnings and implications. *Journal of Occupational and Organizational Psychology, 77,* 439–461.

Allen, V., & Levine, J. (1969). Consensus and conformity. *Journal of Experimental Social Psychology, 5,* 389–399.

Amabile, T. (1983). The social psychology of creativity. *Journal of Personality and Social Psychology, 45,* 357–376.

Amabile, T. (1996). *Creativity in context*. Boulder, CO: Westview.

Amason, A. (1996). Distinguishing the effects of functional and dysfunctional conflict on strategic decision making: Resolving a paradox for top management teams. *Academy of Management Journal, 39*(1), 123–148.

Ancona, D., & Caldwell, D. (1990). Information technology and work groups: The case of new product teams. In J. Galegher, R. Kraut, & C. Egido (Eds.), *Intellectual teamwork: Social and technological foundations of cooperative work* (pp. 173–190). Hillsdale, NJ: Lawrence Erlbaum.

Ancona, D., & Caldwell, D. (1992). Demography and design: Predictors of new product team performance. *Organizational Science, 3,* 321–331.

Antonioni, D. (1994). The effects of feedback accountability on upward appraisal ratings. *Personnel Psychology, 47,* 349–356.

Appelbaum, E., & Batt, R. (1994). *The new American workplace*. Ithaca, NY: IRL Press.

Aranda, E., Aranda, L., & Conlon, K. (1998). *Teams: Structure, process, culture, and politics*. Upper Saddle River, NJ: Prentice Hall.

Argyris, C. (1998). Empowerment: The emperor's new clothes. *Harvard Business Review, 76*(3), 98–105.

Armstrong, D., & Cole, P. (1995). Managing distances and differences in geographically distributed workgroups. In S. Jackson & M. Ruderman (Eds.),

Diversity in work teams: Research paradigms for a changing workplace (pp. 187–215). Washington, DC: American Psychological Association.

Arthur, W., Bennett, W., Edens, P., & Bell, S. (2003). Effectiveness of training in organizations: A meta-analysis of design and evaluation features. *Journal of Applied Psychology, 88*(2), 234–245.

Asch, S. (1955, Winter). Opinions and social pressure. *Scientific American*, 31–35.

Axelrod, R. (1984). *The evolution of cooperation.* New York: Basic Books.

Axley, S. (1996). *Communication at work: Management and the communication-intensive organization.* Westport, CT: Quorum Books.

Bales, R. (1966). The equilibrium problem in small groups. In A. Hare, E. Borgatta, & R. Bales (Eds.), *Small groups: Studies in social interactions* (pp. 444–476). New York: Knopf.

Bandura, A. (2000). Exercise of human agency through collective efficacy. *Current Directions in Psychological Science, 9*, 17–20.

Bass, B. (1985). *Leadership and performance beyond expectations.* New York: Free Press.

Battaglia, B. (1992). Skills for managing multicultural teams. *Cultural Diversity at Work, 4*, 4–12.

Baugh, S., & Graen, G. (1997). Effects of team gender and racial composition on perceptions of team performance in cross-functional teams. *Group and Organization Management, 22*, 366–384.

Beal, D., Cohen, R., Burke, M., & McLendon, C. (2003). Cohesion and performance in groups: A meta-analytic clarification of construct relations. *Journal of Applied Psychology, 88(6)*, 989–1004.

Beebe, S., & Masterson, J. (1994). *Communicating in small groups.* New York: HarperCollins.

Beersma, B., Hollenbeck, J., Humphrey, S., Moon, H., Conlon, D., & Ilgen, D. (2003). Cooperation, competition, and team performance: Toward a contingency approach. *Academy of Management Journal, 46(5)*, 572–590.

Belbin, R. (1981). *Team roles at work.* Oxford, UK: Butterworth Heinemann.

Benne, K., & Sheats, P. (1948). Functional group members. *Journal of Social Issues, 4*, 41–49.

Bennis, W., & Biederman, P. (1997). *Organizing genius: The secrets of creative collaboration.* Reading, MA: Addison-Wesley.

Bikson, T., Cohen, S., & Mankin, D. (1999). Information technology and high-performance teams. In E. Sundstom (Ed.), *Supporting work team effectiveness* (pp. 215–245). San Francisco: Jossey-Bass.

Blake, R., & Mouton, J. (1969). *Building a dynamic corporation through grid organizational development.* Reading, MA: Addison-Wesley.

Brewer, M. (1995). Managing diversity: The role of social identities. In S. Jackson & M. Ruderman (Eds.), *Diversity in work teams: Research paradigms for a changing workplace* (pp. 47–68). Washington, DC: American Psychological Association.

Brown, S. (1996). A meta-analysis and review of organizational research on job involvement. *Psychological Bulletin, 120,* 235–255.

Burgess, R. (1968). Communication networks: An experimental reevaluation. *Journal of Experimental Social Psychology, 4,* 324–327.

Burn, S. (2004). *Groups: Theory and practice.* Belmont, CA: Wadsworth.

Burnstein, E., & Vinokur, A. (1977). Persuasive arguments and social comparison as determinants of attitude polarization. *Journal of Experimental Social Psychology, 13,* 315–332.

Burpitt, W., & Bigoness, W. (1997). Leadership and innovation among teams: The impact of empowerment. *Small Group Research, 28,* 414–423.

Bushe, G. (1988). Cultural contradictions of statistical process control in American manufacturing organizations. *Journal of Management, 14,* 19–31.

Byrne, R. (1984). Overcoming computer phobia. In S. Evans & P. Clarke (Eds.), *The computer culture* (pp. 76–101). Indianapolis, IN: White River Press.

Cacioppo, J., Petty, R., & Morris, K. (1983). Effects of need for cognition on message evaluation, recall, and persuasion. *Journal of Personality and Social Psychology, 45,* 805–818.

Caldwell, B., & Uang, S. (1994). Interactions of situation, social, and technological constraints in information technology use in organizations. In G. Bradley & H. Hendrick (Eds.), *Human factors in organizational design and management* (Vol. 4, pp. 531–536). Amsterdam: Elsevier Science.

Cannon-Bowers, J., & Salas, E. (1998). Team performance and training in complex environments: Recent findings from applied research. *Current Directions in Psychological Science, 7,* 83–87.

Cannon-Bowers, J., Tannenbaum, S., Salas, E., & Volpe, C. (1995). Defining competencies and establishing team training requirements. In R. Guzzo & E. Salas (Eds.), *Team effectiveness and decision making in organizations* (pp. 330–380). San Francisco: Jossey-Bass.

Carnevale, A., Gainer, L., & Meltzer, A. (1990). *Workplace basics: The essential skills employers want.* San Francisco: Jossey-Bass.

Carnevale, A., Gainer, L., & Schulz, E. (1990). *Training the technical work force.* San Francisco: Jossey-Bass.

Carnevale, A., & Stone, S. (1995). *The American mosaic: An in-depth report on the future of diversity at work.* New York: McGraw-Hill.

Carnevale, P. (1986). Strategic choice in mediation. *Negotiation Journal, 2,* 41–56.

Casciaro, T., & Lobo, M. (2005). Competent jerks, lovable fools, and the formation of social networks. *Harvard Business Review, 83*(6), 92–99.

Castore, C., & Murnighan, J. (1978). Determinants of support for group decisions. *Organizational Behavior and Human Performance, 22,* 75–92.

Chaiken, S. (1979). Communicator physical attractiveness and persuasion. *Journal of Personality and Social Psychology, 37,* 1387–1397.

Charan, R., & Useem, J. (2002). Why companies fail. *Fortune, 145*(11), 50–62.

Cheng, J. (1983). Interdependence and coordination in organizations: A role system analysis. *Academy of Management Journal, 26,* 156–162.

Cohen, S., & Bailey, D. (1997). What makes teams work: Group effectiveness research from the shop floor to the executive suite. *Journal of Management, 23*, 239–290.

Cohen, S., Ledford, G., & Spreitzer, G. (1996). A predictive model of self-managing work team effectiveness. *Human Relations, 49*, 643–676.

Cole, R. (1989). *Strategies for learning: Small-group activities in American, Japanese, and Swedish industry.* Berkeley: University of California Press.

Collins, J., & Porras, J. (1994). *Built to last: Successful habits of visionary companies.* New York: HarperCollins.

Cosier, R., & Dalton, D. (1990). Positive effects of conflict: A field assessment. *International Journal of Conflict Management, 1*, 81–92.

Cotton, J. (1993). *Employee involvement.* Newbury Park, CA: Sage.

Cox, T. (1995). The complexity of diversity: Challenges and directions for future research. In S. Jackson & M. Ruderman (Eds.), *Diversity in work teams: Research paradigms for a changing workplace* (pp. 235–245). Washington, DC: American Psychological Association.

Culnan, M., & Markus, M. (1987). Information technologies. In F. Jablin, L. Putnam, K. Roberts, & L. Porter (Eds.), *Handbook of organizational communication: An interdisciplinary perspective* (pp. 420–443). Newbury Park, CA: Sage.

Daft, R., & Lengel, R. (1986). Organizational information requirements, media richness, and structural design. *Management Science, 32*, 554–571.

Dalkey, N. (1969). *The Delphi method: An experimental study of group decisions.* Santa Monica, CA: RAND.

Davis, L., & Wacker, G. (1987). Job design. In G. Salvendy (Ed.), *Handbook of human factors* (pp. 431–452). New York: John Wiley.

Davis, S. (1984). *Managing corporate culture.* Cambridge, MA: Ballinger.

Dawes, R. (1988). *Rational choice in an uncertain world.* San Diego: Harcourt Brace Jovanovich.

Day, D., Gronn, P. & Salas, E. (2004). Leadership capacity in teams. *The Leadership Quarterly, 15(6)*, 857–880.

Deal, T., & Kennedy, A. (1982). *Corporate cultures.* Reading, MA: Addison-Wesley.

Deci, E. (1975). *Intrinsic motivation.* New York: Plenum.

DeDreu, C., & Weingart, L. (2003). Task versus relationship conflict, team performance, and team member satisfaction: A meta-analysis. *Journal of Applied Psychology, 88(4)*, 741–749.

Delbeq, A., Van de Ven, A., & Gustafson, D. (1975). *Group techniques for program planning.* Glenview, IL: Scott, Foresman.

de Leede, J., & Stoker, J. (1999). Self-managing teams in manufacturing companies: Implications for the engineering function. *Engineering Management Journal, 11(3)*, 19–24.

DeMatteo, J., Eby, L., & Sundstrom, E. (1998). Team-based rewards: Current empirical evidence and directions for future research. *Research in Organizational Behavior, 20*, 141–183.

Dennis, A., & Valacich, J. (1993). Computer brainstorms: More heads are better than one. *Journal of Applied Psychology, 78,* 531–537.

Dertouzos, M. (1997). *What will be: How the new world of information will change our lives.* New York: Harper.

Deutsch, M., & Gerard, H. (1955). A study of normative and informational social influence upon individual judgment. *Journal of Abnormal and Social Psychology, 51,* 629–636.

Devine, D., Clayton, L., Philips, J., Dunford, B., & Melner, S. (1999). Teams in organizations: Prevalence, characteristics, and effectiveness. *Small Group Research, 30,* 678–711.

Dewey, J. (1910). *How we think.* New York: Heath.

Diehl, M., & Stroebe, W. (1987). Productivity loss in brainstorming groups: Toward a solution of a riddle. *Journal of Personality and Social Psychology, 53,* 497–509.

DiSalvo, V., Nikkel, E., & Monroe, C. (1989). Theory and practice: A field investigation and identification of group members' perception of problems facing natural work groups. *Small Group Behavior, 20,* 551–567.

Drach-Zahavy, A. (2004). Exploring team support: The role of team's design, values, and leader's support. *Group Dynamics, 8*(4), 235–252.

Driskell, J., Radtke, P. & Salas, E. (2003). Virtual teams: Effects of technological mediation on team performance. *Group Dynamics: Theory, Research, and Practice, 7*(4), 297–323.

Druskat, V., & Wheeler, J. (2003). Managing from the boundary: The effective leadership of self-managing work teams. *Academy of Management Journal, 46*(4), 435–457.

Dyer, W. (1995). *Team building: Current issues and new alternatives.* Reading, MA: Addison-Wesley.

Eagly, A., Karau, S., & Makhijani, M. (1995). Gender and the effectiveness of leaders: A meta-analysis. *Journal of Personality and Social Psychology, 117,* 125–145.

Eagly, A., Wood, W., & Chaiken, S. (1978). Causal inferences about communicators and their effect on opinion change. *Journal of Personality and Social Psychology, 36*(4), 424–435.

Earley, P., & Gibson, C. (2002). *Multinational work teams: A new perspective.* Mahwah, NJ: Lawrence Erlbaum.

Earley, P., & Mosakowski, E. (2000). Creating hybrid team cultures: An empirical test of transnational team functioning. *Academy of Management Journal, 43*(1), 26–49.

Edmondson, A., Bohmer, R., & Pisano, G. (2001). Speeding up team learning. *Harvard Business Review, 79*(9), 125–132.

Edosomwan, J. (1989). *Integrating innovation and technology management.* New York: John Wiley.

Eisenstat, R., & Cohen, S. (1990). Summary: Top management groups. In R. Hackman (Ed.), *Groups that work (and those that don't)* (p. 78). San Francisco: Jossey Bass.

Ellis, L., & Honig-Haftel, S. (1992, March). Reward strategies for R&D. *Research Technology Management*, 16–20.

Falbe, C., & Yukl, G. (1992). Consequences for managers using single influence tactics and combination of tactics. *Academy of Management Journal, 35*, 638–652.

Fan, E., & Gruenfeld, D. (1998). When needs outweigh desires: The effects of resource interdependence and reward interdependence on group problem solving. *Basic and Applied Social Psychology, 20*(1), 45–56.

Farmer, S., & Roth, J. (1998). Conflict-handling behavior in work groups: Effects of group structure, decision process, and time. *Small Group Research, 29*, 669–713.

Feldman, D. (1984). The development and enforcement of group norms. *Academy of Management Review, 9*, 47–53.

Finegan, J. (1993, July). People power. *Inc.*, 62–63.

Fiore, S., & Salas, E. (2004). Why we need team cognition. In E. Salas & S. Fiore (Eds.) *Team cognition: Understanding the factors that drive process and performance* (pp. 235–248). Washington, DC: American Psychological Association.

Fiore, S., & Schooler, J. (2004). Process mapping and shared cognition: Teamwork and the development of shared problem models. In E. Salas & S. Fiore (Eds.) *Team cognition: Understanding the factors that drive process and performance* (pp. 133–152). Washington, DC: American Psychological Association.

Fisher, R., Ury, W., & Patton, B. (1991). *Getting to yes: Negotiating agreement without giving in* (2nd ed.). Boston: Houghton Mifflin.

Fodor, E. (1976). Group stress, authoritarian style of control, and use of power. *Journal of Applied Psychology, 61*, 313–318.

Ford, R., & Fottler, M. (1995). Empowerment: A matter of degree. *Academy of Management Executive, 9*(3), 21–31.

Forsyth, D. (1999). *Group dynamics* (3rd ed.). Belmont, CA: Thompson.

Forsyth, D., & Kelley, K. (1996). Heuristic-based biases in estimates of personal contributions to collective endeavors. In J. Nye & A. Brower (Eds.), *What's social about social cognition? Research on socially shared cognitions in small groups* (pp. 106–123). Thousand Oaks, CA: Sage.

Franz, C., & Jin, K. (1995). The structure of group conflict in a collaborative work group during information systems development. *Journal of Applied Communication Research, 23*, 108–127.

Franz, R. (1998). Task interdependence and personal power in teams. *Small Group Research, 29*, 226–253.

Franz, C., & Jin, K. (1995). The structure of group conflict in a collaborative work group during information systems development. *Journal of Applied Communication Research, 23*, 108–127.

French, J., & Raven, B. (1959). The bases of power. In D. Cartwright (Ed.), *Studies in social power* (pp. 150–167). Ann Arbor: University of Michigan Press.

French, W., & Bell, C. (1984). *Organizational development: Behavioral science interventions for organization improvement (3rd ed.).* Englewood Cliffs, NJ: Prentice Hall.

Gallaway, G. (1996). Facilitating the utilization of information resources in an organization through a sociotechnical systems approach. In O. Brown & H. Hendrick (Eds.), *Human factors in organizational design and management* (5th ed.) (pp. 95–100). Amsterdam: Elsevier Science.

Gandz, J., & Murray, V. (1980). The experience of workplace politics. *Academy of Management Journal, 23,* 237–251.

Gardenswartz, L., & Rowe, A. (1994). *The managing diversity survival guide.* Burr Ridge, IL: Irwin.

Gersick, C. (1988). Time and transition in work teams: Toward a new model of group development. *Academy of Management Journal, 31,* 9–41.

Gersick, C., & Davis-Sacks, M. (1990). Summary: Task forces. In R. Hackman (Ed.), *Groups that work (and those that don't)* (pp. 146–153). San Francisco: Jossey-Bass.

Gibb, J. (1961). Defensive communication. *Journal of Communication, 11,* 141–148.

Gigone, D., & Hastie, R. (1997). The impact of information on small group choice. *Journal of Personality and Social Psychology, 72,* 132–140.

Gilovich, T., Savitsky, K., & Medvec, V. (1998). The illusion of transparency: Biased assessments of others' ability to read one's emotional states. *Journal of Personality and Social Psychology, 75,* 332–346.

Goethals, G., & Nelson, E. (1973). Similarity in the influence process: The belief-value distinction. *Journal of Personality and Social Psychology, 25,* 117–122.

Goldstein, I., & Ford, J. (2002). *Training in organizations (*4th ed.*).* Belmont, CA: Wadsworth.

Graen, G., & Uhl-Bien, M. (1995). Relationship-based approach to leadership: Development of leader-member exchange (LMX) theory of leadership over 25 years. *Leadership Quarterly, 6,* 219–247.

Greenberg, J., & Baron, R. (1997). *Behavior in organizations: Understanding the human side of work* (6th ed.). Upper Saddle River, NJ: Prentice Hall.

Gross, S. (1995). *Compensation for teams: How to design and implement team-based reward programs.* New York: Amacom.

Guzzo, R., & Dickson, M. (1996). Teams in organizations: Recent research on performance and effectiveness. *Annual Review of Psychology, 47,* 307–338.

Gwynne, S. (1990, October 29). The right stuff. *Time,* 74–84.

Hackett, D., & Martin, C. (1993). *Facilitation skills for team leaders.* Menlo Park, CA: CRISP Publications.

Hackman, R. (1986). The psychology of self-management in organizations. In M. Pallak & R. Perloff (Eds.), *Psychology and work* (pp. 89–136). Washington, DC: American Psychological Association.

Hackman, R. (1987). The design of work teams. In J. Lorsch (Ed.), *Handbook of organizational behavior* (pp. 315–342). Englewood Cliffs, NJ: Prentice Hall.

Hackman, R. (1990a). Creating more effective work groups in organizations. In R. Hackman (Ed.), *Groups that work (and those that don't)* (pp. 479–504). San Francisco: Jossey-Bass.

Hackman, R. (1990b). Work teams in organizations: An orienting framework. In R. Hackman (Ed.), *Groups that work (and those that don't)* (pp. 1–14). San Francisco: Jossey-Bass.

Hackman, R. (1992). Group influences on individuals in organizations. In M. Dunnette & L. Hough (Eds.), *Handbook of industrial and organizational psychology* (pp. 199–267). Palo Alto, CA: Consulting Psychologists Press.

Hackman, R. (2002). *Leading teams: Setting the stage for great performances.* Boston: Harvard Business School Press.

Hackman, R., & Morris, C. (1975). Group tasks, group interaction process, and group performance effectiveness: A review and proposed integration. *Advances in Experimental Social Psychology, 8,* 47–99.

Hackman, R., & Oldham, G. (1980). *Work redesign.* Reading, MA: Addison-Wesley.

Hackman, R., & Wageman, R. (2005). A theory of team coaching. *Academy of Management Review, 30*(2), 269–287.

Hackman, R., & Walton, R. (1986). Leading groups in organizations. In P. Goodman (Ed.), *Designing effective work groups* (pp. 72–119). San Francisco: Jossey-Bass.

Hare, A. (1982). *Creativity in small groups.* Beverly Hills, CA: Sage.

Harkins, S., & Jackson, J. (1985). The role of evaluation in eliminating social loafing. *Personality and Social Psychology Bulletin, 11,* 457–465.

Harris, T. (1993). *Applied organizational communication: Perspectives, principles, and pragmatics.* Hillsdale, NJ: Lawrence Erlbaum.

Harrison, D., Price, K., Gavin, J. & Florey, A. (2002). Time, teams, and task performance: Changing effects of surface and deep level diversity on group functioning. *Academy of Management Journal, 45*(5), 1029–1045.

Harvey, J. (1988). *The Abilene paradox and other meditations on management.* Lexington, MA: Lexington Books.

Hayes, N. (1997). *Successful team management.* London: International Thomson Business Press.

Hecht, T., Allen, N., Klammer, J., & Kelly, E. (2002). Group beliefs, ability and performance: The potency of group potency. *Group Dynamics: Theory, Research and Practice, 6*(2), 143–152.

Hemphill, J. (1961). Why people attempt to lead. In L. Petrullo & B. Bass (Eds.), *Leadership and interpersonal behavior* (pp. 201–215). New York: Holt, Rinehart & Winston.

Herrenkohl, R. (2004). *Becoming a team.* Mason, OH: South-Western.

Herrenkohl, R., Judson, G., & Heffner, J. (1999). Defining and measuring employee empowerment. *Journal of Applied Behavioral Science, 35,* 373–389.

Hersey, P., & Blanchard, K. (1993). *Management of organizational behavior: Utilizing human resources.* Englewood Cliffs, NJ: Prentice Hall.

Hertel, G., Geister, S., & Konradt, U. (2005). Managing virtual teams: A review of current empirical research. *Human Resource Management Review, 15(1), 69–95.*

Higgins, E. (1999). Saying is believing effects: When sharing reality about something biases knowledge and evaluations. In L. Thompson, J. Levine, & D. Messick (Eds.), *Shared cognition in organizations: The management of knowledge.* Mahwah, NJ: Lawrence Erlbaum.

Hitrop, J. (1989). Factors associated with successful labor mediation. In K. Kressel & D. Pruitt (Eds.), *Mediation research* (pp. 241–262). San Francisco: Jossey-Bass.

Hofstede, G. (1980). *Culture's consequences: International differences in work related value.* Beverly Hills, CA: Sage.

Hogg, M. (1992). *The social psychology of group cohesiveness: From attraction to social identity.* New York: New York University Press.

Hollander, E., & Offerman, L. (1990). Power and leadership in organizations. *American Psychologist, 45,* 179–189.

Hurwitz, A., Zander, A., & Hymovitch, B. (1953). Some effects of power on the relations among group members. In D. Cartwright & A. Zander (Eds.), *Group dynamics: Research and theory* (pp. 483–492). New York: Harper & Row.

Ilgen, D., Hollenbeck, J., Johnson, M., & Jundt, D. (2005). Teams in organizations: From input-process-output models to IMOI models. *Annual Review of Psychology, 56,* 517–543.

Insko, C., Schopler, J., Graetz, K., Drigotas, S., Currey, K., Smith, S., Brazil, D., & Bornstein, G. (1994). Interindividual-intergroup discontinuity in the prisoner's dilemma game. *Journal of Conflict Resolution, 38,* 87–116.

Jackson, A., & Ruderman, M. (1995). Introduction: Perspective for understanding diverse work teams. In S. Jackson & M. Ruderman (Eds.), *Diversity in work teams: Research paradigms for a changing workplace* (pp. 1–13). Washington, DC: American Psychological Association.

Jackson, S. (1992). Team composition in organizational settings: Issues in managing an increasingly diverse workforce. In S. Worchel, W. Wood, & J. Simpson (Eds.), *Group process and productivity* (pp. 138–173). Newbury Park, CA: Sage.

Janis, I. (1972). *Victims of groupthink.* Boston: Houghton Mifflin.

Janis, I., & Mann, L. (1977). *Decision making.* New York: Free Press.

Janz, B., Colquitt, J., & Noe, R. (1997). Knowledge worker team effectiveness: The role of autonomy, interdependence, team development, and contextual support variables. *Personnel Psychology, 50,* 877–905.

Jehn, K. (1995). A multimethod examination of the benefits and detriments of intragroup conflict. *Administrative Science Quarterly, 40,* 256–282.

Jehn, K., & Shaw, P. (1997). Interpersonal relationships and task performance: An examination of mediating processes in friendship and acquaintance groups. *Journal of Personality and Social Psychology, 72,* 775–790.

Jentsch, F., & Smith-Jentsch, K. (2001). Assertiveness and team performance: More than "just say no". In E. Salas, C. Bowers, & E. Edens (Eds.), *Improving teamwork in organizations* (pp. 73–94). Mahwah, NJ: Lawrence Erlbaum.

Johnson, D., & Johnson, F. (1997). *Joining together: Group theory and group skills* (6th ed.). Boston: Allyn & Bacon.

Johnson, D., Maruyama, G., Johnson, R., Nelson, D., & Skon, L. (1981). Effects of cooperative, competitive, and individualistic goal structures on achievement: A meta-analysis. *Psychological Bulletin, 89,* 47–62.

Jones, G., & George, J. (1998). The experience and evolution of trust: Implications for cooperation and teamwork. *Academy of Management Review, 23,* 531–546.

Jones, R., & Brehm, J. (1970). Persuasiveness of one and two sided communications as a function of awareness when there are two sides. *Journal of Experimental Social Psychology, 6,* 47–56.

Jones, S., & Moffett, R. (1999). Measurement and feedback systems for teams. In E. Sundstom (Ed.), *Supporting work team effectiveness* (pp. 157–187). San Francisco: Jossey-Bass.

Kanter, R. (1977). *Men and women of the corporation.* New York: Basic Books.

Karau, S., & Williams, K. (1993). Social loafing: A meta-analytic review and theoretical integration. *Journal of Personality and Social Psychology, 65,* 681–706.

Karau, S., & Williams, K. (1997). The effects of group cohesion on social loafing and social compensation. *Group Dynamics: Theory, Research, and Practice, 1,* 156–168.

Katzenbach, J., & Smith, D. (1993). *The wisdom of teams.* Cambridge, MA: Harvard Business School Press.

Katzenbach, J., & Smith, D. (2001). *The discipline of teams.* New York: John Wiley & Sons.

Kayser, T. (1990). *Mining group gold.* El Segundo, CA: Sherif Publishing.

Kemery, E., Bedeian, A., Mossholder, K., & Touliatos, J. (1985). Outcomes of role stress: A multisample constructive replication. *Academy of Management Review, 28,* 363–375.

Kerr, N., & Bruun, S. (1983). Dispensability of member effort and group motivation losses: Free rider effects. *Journal of Personality and Social Psychology, 44,* 78–94.

Kerr, N., & Tindale, R. (2004). Group performance and decision making. *Annual Review of Psychology, 55,* 623–655.

Keysar, B. (1998). Language users as problem solvers: Just what ambiguity problem do they solve? In S. Fussell & R. Kreuz (Eds.), *Social and cognitive approaches to interpersonal communication* (pp. 175–200). Mahwah, NJ: Lawrence Erlbaum.

Keysar, B. & Henly, A. (2002). Speakers' overestimation of their effectiveness. *Psychological Science, 13(3),* 207–212.

Kidder, T. (1981). *The soul of the new machine.* New York: Avon.

Kiesler, S. (1986, January). The hidden messages in computer networks. *Harvard Business Review,* 46–60.

Kiesler, S., Siegel, J., & McGuire, T. (1984). Social aspects of computer mediated communication. *American Psychologist, 39,* 1123–1134.

Kilmann, R., & Saxton, M. (1983). *The Kilmann-Saxton culture gap survey.* Pittsburgh, PA: Organizational Design Consultants.

Kipnis, D. (1976). *The powerholders.* Chicago: University of Chicago Press.

Kipnis, D., & Schmidt, S. (1982). *Profiles of organizational influence strategies: Influencing your subordinates.* San Diego: University Associates.

Kipnis, D., Schmidt, S., Swaffin-Smith, C., & Wilkinson, I. (1984). Patterns of managerial influence: Shotgun managers, tacticians, and bystanders. *Organizational Dynamics, 12*(3), 58–67.

Kirkman, B., & Rosen, B. (1999). Beyond self-management: Antecedents and consequences of team empowerment. *Academy of Management Journal, 42*(1), 58–74.

Kirkpatrick, S., & Locke, E. (1991). Leadership: Do traits matter? *Academy of Management Executive, 5,* 48–60.

Kivlighan, D., & Jauquet, C. (1990). Quality of group member agendas and group session climate. *Small Group Research, 1,* 205–219.

Klein, J. (1984, September). Why supervisors resist employee involvement. *Harvard Business Review,* 87–93.

Klimoski, R., & Mohammed, S. (1997). Team mental model: Construct or metaphor? *Journal of Management, 20(2),* 403–437.

Knight, G., & Dubro, A. (1984). Cooperative, competitive, and individualistic social values. *Journal of Personality and Social Psychology, 46,* 98–105.

Krauss, R., & Fussell, S. (1991). Perspective-taking in communication: Representations of other's knowledge in reference. *Social Cognition, 9,* 2–24.

Kruger, J., Epley, N., Parker, J., & Ng, Z. (2005). Egocentrism over e-mail: Can we communicate as well as we think? *Journal of Personality and Social Psychology, 89*(6), 925–936.

Langfred, C. (2000). Work group design and autonomy: A field study of the interaction between task interdependence and group autonomy. *Small Group Research, 31,* 54–70.

Larson, C., & LaFasto, F. (1989). *Teamwork: What must go right/what can go wrong.* Newbury Park, CA: Sage.

Larson, J., Foster-Fishman, P., & Franz, T. (1998). Leadership style and discussion of shared and unshared information in decision making groups. *Personality and Social Psychology Bulletin, 24,* 482–495.

Latane, B., Williams, K., & Harkins, S. (1979). Many hands make light the work: The causes and consequences of social loafing. *Journal of Personality and Social Psychology, 37,* 822–832.

Laughlin, P., & Hollingshead, A. (1995). A theory of collective induction. *Organizational Behavior and Human Decision Processes, 61,* 94–107.

Lawler, E. (1986). *High involvement management.* San Francisco: Jossey-Bass.

Lawler, E. (1999). Creating effective pay systems for teams. In E. Sundstom (Ed.), *Supporting work team effectiveness* (pp. 188–214). San Francisco: Jossey-Bass.

Lawler, E. (2000). *Rewarding excellence: Pay strategies for the new economy.* San Francisco: Jossey-Bass.

Lawler, E., & Mohrman, S. (1985, January). Quality circles after the fad. *Harvard Business Review*, 85–71.

Lawler, E., Mohrman, S., & Ledford, G. (1995). *Creating high performance organizations: Practices and results of employee involvement and quality management in Fortune 1000 companies.* San Francisco: Jossey-Bass.

Lea, D., & Brostrom, L. (1988). Managing the high-tech professional. *Personnel*, 65(6), 12–22.

Levi, D. (1988). *The role of corporate culture in selecting human resources policies.* Unpublished manuscript, Northern Telecom.

Levi, D., & Cadiz, D. (1998). *Evaluating teamwork on student projects: The use of behaviorally anchored scales to evaluate student performance.* East Lansing, MI: National Center for Research on Teacher Learning. (ERIC Document Reproduction Service No. TM029122).

Levi, D., & Lawn, M. (1993). The driving and restraining forces which affect technological innovation. *Journal of High Technology Management Research, 4*, 225–240.

Levi, D., & Rinzel, L. (1998). Employee attitudes toward various communications technologies when used for communicating about organizational change. In P. Vink, E. Koningsveld, & S. Dhondt (Eds.), *Human factors in organizational design and management* (Vol. 6, pp. 483–488). Amsterdam: Elsevier Science.

Levi, D., & Slem, C. (1990). *The human, social, and organizational impact of electronic mail.* Unpublished manuscript, Institute for Information Studies, Falls Church, VA.

Levi, D., & Slem, C. (1995). Team work in research and development organizations: The characteristics of successful teams. *International Journal of Industrial Ergonomics, 16*, 29–42.

Levi, D., & Slem, C. (1996). The relationship of concurrent engineering practices to different views of project success. In O. Brown & H. Hendrick (Eds.), *Human factors in organizational design and management* (Vol. 5, pp. 25–30). Amsterdam: Elsevier Science.

Levine, J. (1989). Reaction to opinion deviance in small groups. In P. Paulus (Ed.), *Psychology of group influence: New perspectives* (pp. 187–232). Hillsdale, NJ: Lawrence Erlbaum.

Levine, J., & Choi, H. (2004). Impact of personnel turnover on team performance and cognition. In E. Salas & S. Fiore (Eds.) *Team cognition: Understanding the factors that drive process and performance* (pp. 153–175). Washington, DC: American Psychological Association.

Lewin, K. (1951). *Field theory in social science.* New York: Harper.

Likert, R. (1961). *New patterns in management.* New York: McGraw-Hill.

Locke, E., & Latham, G. (1990). *A theory of goal setting and task performance.* Englewood Cliffs, NJ: Prentice Hall.

Lord, R. (1985). An information processing approach to social perceptions, leadership, and behavioral measurement. *Research in Organizational Behavior, 7*, 87–128.

Lott, A., & Lott, B. (1965). Group cohesiveness as interpersonal attraction: A review of the relationships with antecedent and consequence variables. *Psychological Bulletin, 64*, 259–309.

Lumsden, G., & Lumsden, D. (1997). *Communicating in groups and teams*. Belmont, CA: Wadsworth.

Luthans, F., & Fox, M. (1989, March). Update on skill-based pay. *Personnel*, 26–31.

Mannix, E., & Klimoski, R. (2005). What differences make a difference? The promise and reality of diverse teams in organizations. *Psychology in the Public Interest, 6*(2), 32–55.

Manufacturing Studies Board. (1986). *Human resources practices for implementing advanced manufacturing technology*. Washington, DC: National Academy Press.

Manz, C. (1992). Self-leading work teams: Moving beyond self–management myths. *Human Relations, 45*, 1119–1140.

Marks, M., Mathieu, J., & Zaccaro, S. (2001). A temporally based framework and taxonomy of team processes. *Academy of Management Review, 26*(3), 356–376.

Marks, M., Sabella, M., Burke, C., & Zaccaro, S. (2002). The impact of cross-training on team effectiveness. *Journal of Applied Psychology, 87*(1), 2–13.

Marquardt, M. (2002). *Building the learning organization*. Palo Alto, CA: Davies-Black.

Marquardt, M. (2004, June). Harnessing the power of Action Learning. *Training & Development*, 26–32.

Mayo, E. (1933). *The human problems of an industrial civilization*. Cambridge, MA: Harvard University Press.

McAllister, D. (1995). Affect and cognition based trust as foundations for interpersonal cooperation in organizations. *Academy of Management Journal, 38*, 24–59.

McClelland, D., & Boyatzis, R. (1982). Leadership motive pattern and long-term success in management. *Journal of Applied Psychology, 67*, 737–743.

McComb, S., Green, S., & Compton, W. (1999). Project goals, team performance, and shared understanding. *Engineering Management Journal, 11*(3), 7–12.

McGrath, J. (1984). *Groups: Interaction and performance*. Englewood Cliffs, NJ: Prentice Hall.

McGrath, J. (1990). Time matters in groups. In J. Galegher, R. Kraut, & C. Egido (Eds.), *Intellectual teamwork: Social and technological foundations of cooperative work* (pp. 23–62). Hillsdale, NJ: Lawrence Erlbaum.

McGrath, J., Berdahl, J., & Arrow, H. (1995). Traits, expectations, culture, and clout: The dynamics of diversity in work groups. In S. Jackson & M. Ruderman (Eds.), *Diversity in work teams: Research paradigms for a changing workplace* (pp. 17–45). Washington, DC: American Psychological Association.

McGrath, J., & Hollingshead, A. (1994). *Groups interacting with technology*. Thousand Oaks, CA: Sage.

McGregor, D. (1960). *The human side of enterprise*. New York: McGraw-Hill.

McGrew, J., Bilotta, J., & Deeney, J. (1999). Software team formation and decay. *Small Group Research, 30*, 209–234.

McIntyre, R., & Salas, E. (1995). Measuring and managing for team performance: Lessons from complex environments. In R. Guzzo & E. Salas (Eds.), *Team effectiveness and decision making in organizations* (pp. 9–45). San Francisco: Jossey-Bass.

McKenna, E. (1994). *Business psychology and organizational behavior.* Hillsdale, NJ: Lawrence Erlbaum.

Meindl, J., & Ehrlich, S. (1987). The romance of leadership and the evaluation of organizational performance. *Academy of Management Journal, 30,* 91–109.

Milgram, S. (1974). *Obedience to authority.* New York: Harper & Row.

Mittleman, D., & Briggs, R. (1999). Communication technologies for traditional and virtual teams. In E. Sundstom (Ed.), *Supporting work team effectiveness* (pp. 246–270). San Francisco: Jossey-Bass.

Mohrman, S. (1993). Integrating roles and structure in the lateral organization. In J. Galbraith & E. Lawler (Eds.), *Organizing for the future* (pp. 109–141). San Francisco: Jossey-Bass.

Mohrman, S., Cohen, S., & Mohrman, A. (1995). *Designing team-based organizations.* San Francisco: Jossey-Bass.

Moreland, R., Argote, L., & Krishnan, R. (1996). Socially shared cognition at work. In J. Nye & A. Bower (Eds.), *What's social about social cognition?* Thousand Oaks, CA: Sage.

Moreland, R., & Levine, J. (1982). Socialization in small groups: Temporal changes in individual-group relations. *Advances in Experimental Social Psychology, 15,* 137–192.

Moreland, R., & Levine, J. (1989). Newcomers and old-timers in small groups. In P. Paulus (Ed.), *Psychology of group influence* (pp. 143–186). Hillsdale, NJ: Lawrence Erlbaum.

Moreland, R., & Levine, J. (1992). Problem identification by groups. In S. Worchel, W. Wood, & J. Simpson (Eds.), *Group process and productivity* (pp. 17–48). Newbury Park, CA: Sage.

Morgeson, F. (2005). The external leadership of self-managing teams: Intervening in the context of novel and disruptive events. *Journal of Applied Psychology, 90(3),* 497–508.

Moscovici, S. (1985). Social influence and conformity. In G. Lindzey & E. Aronson (Eds.), *The handbook of social psychology* (pp. 347–412). Hillsdale, NJ: Lawrence Erlbaum.

Mullen, B., & Copper, C. (1994). The relation between group cohesiveness and performance: An integration. *Psychological Bulletin, 115,* 210–227.

Mullen, B., Johnson, C., & Salas, E. (1991). Productivity loss in brainstorming groups: A meta-analytic integration. *Basic and Applied Psychology, 12,* 3–24.

Mullen, B., Salas, E., & Driskell, J. (1989). Salience, motivation, and artifacts as contributors to the relationship between participation rate and leadership. *Journal of Experimental Social Psychology, 25,* 545–559.

Murnighan, J. (1981). Group decision making: What strategies should you use? *Management Review, 25,* 56–62.

Myers, D., & Lamm, H. (1976). The group polarization phenomenon. *Psychological Bulletin, 83,* 602–627.

Nadler, J., Thompson, L., & Morris, M. (1999, August). *Schmooze or lose: The efforts of rapport and gender in e-mail negotiations.* Paper presented at the annual meeting of the Academy of Management, Chicago.

Naquin, C., & Tynan, R. (2003) The team halo effect: Why teams are not blamed for their failures. *Journal of Applied Psychology, 88(2),* 332–340.

Nemeth, C. (1979). The role of an active minority in intergroup relations. In W. Austin & S. Worchel (Eds.), *The social psychology of intergroup relations* (p. 348-362). Pacific Grove, CA: Brooks/Cole.

Nemeth, C. (1997). Managing innovation: When less is more. *California Management Review, 40*(1), 58–66.

Nemeth, C., & Staw, B. (1989). The trade-offs of social control and innovation in groups and organizations. In L. Berkowitz (Ed.), *Advances in experimental social psychology* (pp. 195–230). San Diego: Academic Press.

Nkomo, S. (1995). Identities and the complexity of diversity. In S. Jackson & M. Ruderman (Eds.), *Diversity in work teams: Research paradigms for a changing workplace* (pp. 247–253). Washington, DC: American Psychological Association.

Northcraft, G., Polzer, J., Neale, M., & Kramer, R. (1995). Diversity, social identity, and performance: Emergent social dynamics in cross-functional teams. In S. Jackson & M. Ruderman (Eds.), *Diversity in work teams: Research paradigms for a changing workplace* (pp. 69–95). Washington, DC: American Psychological Association.

Nye, J., & Forsyth, D. (1991). The effects of prototype-based biases on leadership appraisals: A test of leadership categorization theory. *Small Group Research, 22,* 360–379.

O'Dell, C. (1989, November 1). Team play, team pay: New ways of keeping score. *Across the Board,* 38–45.

Offner, A., Kramer, T., & Winter, J. (1996). The effects of facilitation, recording, and pauses on group brainstorming. *Small Group Research, 27,* 283–298.

Orpen, C. (1979). The effects of job enrichment on employee satisfaction, motivation, involvement, and performance: A field experiment. *Human Relations, 32,* 189–217.

Orsburn, J., Moran, L., Musselwhite, E., Zenger, J., & Perrin, C. (1990). *Self-directed work teams: The new American challenge.* Homewood, IL: Business One Irwin.

Osborn, A. (1957). *Applied imagination.* New York: Scribner.

Osgood, C. (1962). *An alternative to war and surrender.* Urbana: University of Illinois Press.

Ouchi, W. (1981). *Theory Z: How American business can meet the Japanese challenge.* Reading, MA: Addison-Wesley.

Parks, C. (1994). The predictive ability of social values in resource dilemmas and public good games. *Personality and Social Psychology Bulletin, 20,* 431–438.

Parks, C., & Sanna, L. (1999). *Group performance and interaction.* Boulder, CO: Westview.

Pascale, R., & Athos, A. (1981). *The art of Japanese management.* New York: Simon & Schuster.

Paulus, P. (1998). Developing consensus about groupthink after all these years. *Organization Behavior and Human Decision Processes, 73,* 362–374.

Paulus, P. (2000). Groups, teams, and creativity: The creative potential of idea-generating groups. *Applied Psychology: An International Review, 49*(2), 237–262.

Paulus, P. (2002). Different ponds for different fish: A contrasting perspective on team innovation. *International Association for Applied Psychology,* 394–399.

Pavit, C. (1993). What (little) we know about formal group discussion procedures. *Small Group Research, 24,* 217–235.

Pelled, L., Eisenhardt, K., & Xin, K. (1999). Exploring the black box: An analysis of work group diversity, conflict and performance. *Administrative Science Quarterly, 44,* 1–28.

Peters, T., & Waterman, R. (1982). *In search of excellence.* New York: Harper & Row.

Peterson, R., & Nemeth, C. (1996). Focus versus flexibility: Majority and minority influence can both improve performance. *Personality and Social Psychology Bulletin, 22,* 14–24.

Podsakoff, P., & Schriesheim, C. (1985). Field studies of French and Raven's bases of power. *Psychological Bulletin, 97,* 387–411.

Pokras, S. (1995). *Team problem solving.* Menlo Park, CA: CRISP Publications.

Poole, M. (1983). Decision development in small groups: A multiple sequence model of group decision development. *Communication Monographs, 50,* 321–330.

Prochaska, R. (1980). The management of innovation in Japan: Why it is successful. *Research Management, 23,* 35–38.

Pruitt, D. (1981). *Negotiation behavior.* New York: Academic Press.

Pruitt, D. (1986). Trends in the scientific study of negotiation. *Negotiation Journal, 2,* 237–244.

Pruitt, D., & Carnevale, P. (1993). *Negotiation in social conflict.* Pacific Grove, CA: Brooks/Cole.

Rahim, M. (1983). A measure of styles of handling interpersonal conflict. *Academy of Management Journal, 26,* 368–376.

Raven, B., Schwarzwald, J., & Koslowsky, M. (1998). Conceptualizing and measuring a power/interaction model of interpersonal influence. *Journal of Applied Social Psychology, 28,* 307–333.

Reichwald, R., & Goecke, R. (1994). New communication media and new forms of cooperation in the top management area. In G. Bradley & H. Hendrick (Eds.), *Human factors in organizational design and management* (Vol. 4, pp. 511–518). Amsterdam: Elsevier Science.

Rice, R., Instone, D., & Adams, J. (1984). Leader sex, leader success, and leadership process: Two field studies. *Journal of Applied Psychology, 69,* 12–31.

Robbins, S. (1974). *Managing organizational conflict.* Englewood Cliffs, NJ: Prentice Hall.

Roch, S., & Ayman, R. (2005). Group decision making and perceived decision support: The role of communication medium. *Group Dynamics, 9*(1), 15–31.

Roethlisberger, F., & Dickson, W. (1939). *Management and the worker.* Cambridge, MA: Harvard University Press.

Rohlen, T. (1975). The company work group. In E. Vogel (Ed.), *Modern Japanese organization and decision making* (pp. 185–209). Tokyo: Tuttle.

Rosenberg, L. (1961). Group size, prior experience, and conformity. *Abnormal and Social Psychology, 63,* 436–437.

Ross, L., & Ward, A. (1995). Psychological barriers to dispute resolution. In M. Zanna (Ed.), *Advances in experimental social psychology* (Vol. 27, pp. 255–304). San Diego: Academic Press.

Rynes, S., Gerhart, B., & Parks, L. (2005). Personnel psychology: Performance evaluation and pay for performance. *Annual Review of Psychology, 56,* 571–600.

Safizadeh, M. (1991). The case of workgroups in manufacturing operations. *California Management Review, 33* (4), 61–82.

Sakuri, M. (1975). Small group cohesiveness and detrimental conformity. *Sociometry, 38,* 340–357.

Salas, E., Bowers, C., & Edens, E. (2001). Research and practice of resource management in organizations. In E. Salas, C. Bowers, & E. Edens (Eds.), *Improving teamwork in organizations: Applications of resource management training* (pp. 235–240). Mahwah, NJ: Lawrence Erlbaum.

Salas, E., & Cannon-Bowers, J. (2001). The science of training: A decade of progress. In S. Fiske, D. Schacter, & C. Zahn-Waxler (Eds.), *Annual Review of Psychology* (pp. 471–499). Palo Alto, CA: Annual Review Press.

Schein, E. (1988). *Process consultation: Its role in organizational development.* Reading, MA: Addison-Wesley.

Schein, E. (1992). *Organizational culture and leadership* (2nd ed.). San Francisco: Jossey-Bass.

Scholtes, P. (1988). *The team handbook: How to use teams to improve quality.* Madison, WI: Joiner Associates.

Scholtes, P. (1994). *The team handbook for educators.* Madison, WI: Joiner Associates.

Schwenk, C. (1990). Effects of devil's advocacy and dialectical inquiry on decision making: A meta-analysis. *Organizational Behavior and Human Decision Processes, 47,* 161–176.

Shaw, M. (1978). Communication networks fourteen years later. In L. Berkowitz (Ed.), *Group processes* (pp. 351–356). New York: Academic Press.

Shaw, M. (1981). *Group dynamics: The psychology of small group behavior.* New York: McGraw-Hill.

Sherif, M. (1966). *In common predicament: Social psychology of intergroup conflict and cooperation.* Boston: Houghton Mifflin.

Siegel, S., & Fouraker, L. (1960). *Bargaining and group decision making.* New York: McGraw-Hill.

Simon, H. (1979). *The science of the artificial* (2nd ed.). Cambridge: MIT Press.

Slavin, R. (1985). Cooperative learning: Applying contact theory in desegregated schools. *Journal of Social Issues, 41,* 45–62.

Slem, C., Levi, D., & Young, A. (1995). Attitudes about the impact of technological change: Comparison of U.S. and Japanese workers. *Journal of High Technology Management Research, 6,* 211–228.

Smith, K., Carrol, S., & Ashford, S. (1995). Intra- and interorganizational cooperation: Toward a research agenda. *Academy of Management Journal, 38,* 7–23.

Smith-Jentsch, K., Salas, E., & Brannick, M. (2001). To transfer or not to transfer? Investigating the combined effects of trainee characteristics, team leader support, and team climate. *Journal of Applied Psychology, 86*(2), 279–292.

Snell, S., Snow, C., Davison, S., & Hambrick, D. (1998). Designing and supporting transnational teams: The human resource agenda. *Human Resource Management, 37*(2), 147–158.

Snow, C., Snell, S., Davison, S., & Hambrick, D. (1996). Use transnational teams to globalize your company. *Organizational Dynamics, 24*(4), 50–67.

Spreitzer, G., Cohen, S., & Ledford, G. (1999). Developing effective self-managing work teams in service organizations. *Group and Organization Management, 24,* 340–367.

Sproull, L., & Kiesler, S. (1991). *Connections: New ways of working in the networked organization.* Cambridge: MIT Press.

Srull, T., & Wyer, R. (1988). *Advances in social cognition.* Hillsdale, NJ: Lawrence Erlbaum.

Stasser, G. (1992). Pooling of unshared information during group discussions. In S. Worchel, W. Wood, & J. Simpson (Eds.), *Group process and productivity* (pp. 17–48). Newbury Park, CA: Sage.

Stasser, G., & Titus, W. (1985). Pooling of unshared information in group decision making: Biased information sampling during discussion. *Journal of Personality and Social Psychology, 48,* 1467–1478.

Stein, M. (1975). *Stimulating creativity.* New York: Academic Press.

Steiner, I. (1972). *Group process and productivity.* New York: Academic Press.

Stewart, G., & Manz, C. (1995). Leadership for self-managing work teams: A topology and integrative model. *Human Relations, 48,* 747–770.

Stogdill, R. (1974). *Handbook of leadership.* New York: Free Press.

Stoner, J. (1961). *A comparison of individual and group decision making involving risk.* Unpublished master's thesis, Massachusetts Institute of Technology.

Sundstrom, E. (1999a). The challenges of supporting work team effectiveness. In E. Sundstom (Ed.), *Supporting work team effectiveness* (pp. 2–23). San Francisco: Jossey-Bass.

Sundstrom, E. (1999b). Supporting work team effectiveness: Best practices. In E. Sundstom (Ed.), *Supporting work team effectiveness* (pp. 301–342). San Francisco: Jossey-Bass.

Sundstrom, E., DeMeuse, K., & Futrell, D. (1990). Work teams. *American Psychologist, 45,* 120–133.

Sundstrom, E., McIntyre, M., Halfhill, T., & Richards, H. (2000). Work groups: From the Hawthorne studies to work teams of the 1990s and beyond. *Group Dynamics: Theory, Research, and Practice, 4*(1), 44–67.

Sweeney, J. (1973). An experimental investigation of the free rider problem. *Social Science Research, 2,* 277–292.

Taha, L., & Caldwell, B. (1993). Social isolation and integration in electronic environments. *Behaviour and Information Technology, 12,* 276–283.

Tajfel, H. (1982). Social psychology of intergroup relations. *Annual Review of Psychology, 33,* 1–39.

Tajfel, H., & Turner, J. (1986). The social identity theory of intergroup behavior. In S. Worchel & W. Austin (Eds.), *Psychology of intergroup relations* (pp. 2–24). Chicago: Nelson-Hall.

Taylor, F. (1923). *The principles of scientific management.* New York: Harper.

Thomas, B., & Olson, M. (1988). Gain sharing: The design that guarantees success. *Personnel Journal, 67*(5), 73–79.

Thomas, K. (1976). Conflict and conflict management. In M. Dunnette (Ed.), *Handbook of industrial and organizational psychology* (pp. 889–935). Chicago: Rand McNally.

Thomas, K. (1992). Conflict and conflict management: Reflections and update. *Journal of Organizational Behavior, 13,* 265–274.

Thompson, L. (2000). *Making the team: A guide for managers.* Upper Saddle River, NJ: Prentice Hall.

Thompson, L. (2004). *Making the team: A guide for managers* (2nd ed.). Upper Saddle River, NJ: Pearson.

Thompson, L., & Hastie, R. (1990). Judgment tasks and biases in negotiation. In B. Sheppard, M. Bazerman, & R. Lewicki (Eds.), *Research on negotiations in organizations* (Vol. 2, pp. 1077–1092). Greenwich, CT: JAI.

Thompson, L., & Hrebec, D. (1996). Loose-loose agreements in interdependent decision making. *Psychological Bulletin, 120,* 396–409.

Thomsett, R. (1980). *People and project management.* New York: Yourdon Press.

Tjosvold, D. (1995). Cooperation theory, constructive controversy, and effectiveness: Learning from crisis. In R. Guzzo & E. Salas (Eds.), *Team effectiveness and decision making in organizations* (pp. 79–112). San Francisco: Jossey-Bass.

Tolbert, P., Andrews, A., & Simons, T. (1995). The effects of group proportions on group dynamics. In S. Jackson & M. Ruderman (Eds.), *Diversity in work teams: Research paradigms for a changing workplace* (pp. 131–159). Washington, DC: American Psychological Association.

Triandis, H. (1994). *Culture and social behavior.* New York: McGraw-Hill.

Triplett, N. (1898). The dynamogenic factors in pace-making and competition. *American Journal of Psychology, 9,* 507–533.

Tsui, A., Xin, K., & Egan, T. (1995). Relational demography: The missing link in vertical dyad linkage. In S. Jackson & M. Ruderman (Eds.), *Diversity in work teams: Research paradigms for a changing workplace* (pp. 97–129). Washington, DC: American Psychological Association.

Tuckman, B., & Jensen, M. (1977). Stages of small group development revisited. *Group and Organizational Studies, 2,* 419–427.

Uhl-Bien, M., & Graen, G. (1992). Self-management and team-making in cross-functional work teams: Discovering the keys to becoming an integrated team. *Journal of High Technology Management Research, 3,* 225–241.

Uzzi, B. (1997). Social structure and competition in interfirm networks: The paradox of embeddedness. *Administrative Science Quarterly, 42,* 35–67.

Van de Ven, A., & Delbecq, A. (1974). The effectiveness of nominal, Delphi, and interacting group decision making processes. *Academy of Management Journal, 17,* 605–621.

Van der Vegt, G., & Bunderson, J. (2005). Learning and performance in multidisciplinary teams: The importance of collective team identification. *Academy of Management Journal, 48*(3), 532–547.

Van der Vegt, G., Emans, B., & Van de Vliert, E. (1998). Motivating effects of task and outcome interdependence in work teams. *Group and Organization Management, 23,* 124–144.

Van Gundy, A. (1981). *Techniques of structured problem solving.* New York: Van Nostrand Reinhold.

Van Gundy, A. (1987). *Creative problem solving: A guide for trainers and management.* New York: Quorum Books.

Van Maanen, J., & Barley, S. (1985). Cultural organization: Fragments of a theory. In P. Frost (Ed.), *Organizational culture* (pp. 31–54). London: Sage.

Vroom, V., & Jago, A. (1988). *The new leadership: Managing participation in organizations.* Englewood Cliffs, NJ: Prentice Hall.

Vroom, V., & Yetton, P. (1973). *Leadership and decision making.* Pittsburgh, PA: University of Pittsburgh Press.

Walker, H., Ilardi, B., McMahon, A., & Fennell, M. (1996). Gender, interaction, and leadership. *Social Psychology Quarterly, 59,* 255–272.

Wall, V., & Nolan, L. (1987). Small group conflict: A look at equity, satisfaction, and styles of conflict management. *Small Group Behavior, 18,* 188–211.

Walton, R., & Hackman, J. (1986). Groups under contrasting management strategies. In P. Goodman & Associates (Eds.), *Designing effective work groups* (pp. 168–201). San Francisco: Jossey-Bass.

Walton, R., & McKersie, R. (1965). *A behavioral theory of labor negotiations.* New York: McGraw-Hill.

Wanous, J. (1980). *Organizational entry: Recruitment, selection, and socialization of newcomers.* Reading, MA: Addison-Wesley.

Wanous, J., & Youtz, M. (1986). Solution diversity and the quality of group decisions. *Academy of Management Journal, 29,* 149–159.

Watson, R., DeSanctis, G., & Poole, M. (1988). Using a GDSS to facilitate group consensus: Some intended and unintended consequences. *MIS Quarterly, 12,* 463–476.

Wech, B., Mossholder, K., Steel, R., & Bennett, N. (1998). Does work group cohesiveness affect individuals' performance and organizational commitment? *Small Group Research, 29,* 472–494.

Wegner, D. (1986). Transactive memory: A contemporary analysis of the group mind. In B. Mullen & G. Goethals (Eds.), *Theories of group behavior* (pp. 185–208). New York: Springer-Verlag.

Weldon, E., Jehn, K., & Pradhan, P. (1991). Processes that mediate the relationship between a group goal and improved group performance. *Journal of Personality and Social Psychology, 61,* 555–569.

Wellins, R., Byham, W., & Wilson, J. (1991). *Empowered teams.* San Francisco: Jossey-Bass.

Wellins, R., & George, J. (1991). The key to self-directed teams. *Training and Development Journal, 45*(4), 26–31.

Wheelan, S. (2005). *Group process: A developmental perspective.* (2nd ed.). Boston: Allyn & Bacon.

Wilder, D. (1986). Social categorization: Implications for creation and reduction of intergroup bias. In L. Berkowitz (Ed.), *Advances in experimental social psychology* (Vol. 19, pp. 291–355). San Diego: Academic Press.

Wilder, D. (1990). Some determinants of the persuasive power of in-groups and out-groups. *Journal of Personality and Social Psychology, 59,* 1202–1213.

Williams, J., & Best, D. (1990). *Measuring sex differences: A multination study.* Newbury Park, CA: Sage.

Witeman, H. (1991). Group member satisfaction: A conflict-related account. *Small Group Research, 22,* 24–58.

Worchel, S., Andreoli, V., & Folger, R. (1977). Intergroup cooperation and inter-group attraction: The effect of previous interaction and outcome of combined effort. *Journal of Experimental Social Psychology, 13,* 131–140.

Youngs, G. (1986). Patterns of threat and punishment reciprocity in a conflict setting. *Journal of Personality and Social Psychology, 51,* 541–546.

Yukl, G. (1989). Managerial leadership: A review of theory and research. *Journal of Management, 15,* 251–289.

Yukl, G. (1994). *Leadership in organizations* (3rd ed.). Englewood Cliffs, NJ: Prentice Hall.

Yukl, G., & Guinan, P. (1995). Influence tactics used for different objectives with subordinates, peers, and supervisors. *Group and Organization Management, 20,* 272–297.

Zaccaro, S., & Marks, M. (1999). The roles of leaders in high-performance teams. In E. Sundstrom (Ed.), *Supporting work team effectiveness* (pp. 95–125). San Francisco: Jossey-Bass.

Zander, A. (1977). *Groups at work.* San Francisco: Jossey-Bass.

Zander, A. (1994). *Making groups effective.* San Francisco: Jossey-Bass.

Zanna, M. (1993). Message receptivity: A new look at the old problem of open versus closed mindedness. In A. Mitchell (Ed.), *Advertising: Exposure, memory, and choice.* Hillsdale, NJ: Lawrence Erlbaum.

Zarraga, C., & Bonache, J. (2005). The impact of team atmosphere on knowledge outcomes in self-managed teams. *Organizational Studies, 26*(5), 661–681.

Zigon, J. (1997, January-February). Team performance measurement: A process for creating performance standards. *Compensation and Benefits Review*, 38–48.

Zuboff, S. (1988). *In the age of the smart machine*. New York: Basic Books.

Index

expectations approach to, 222
group cohesion and, 230
group creativity and, 207
identity-relevant tasks and, 231
impact of, 229
importance of, 220–221
managing, 232–234
organizational, 222
performance tasks and, 229
power in group and, 226
problem-solving and, 230
problems with. *See* Diversity problems
psychological, 222
socialization and, 228
task assignments and, 234
trait approach to, 222
turnover and, 228
types of, 221–222
Diversity problems:
diversity as cognitive process, 223–225
diversity as social process, 226
emotional distrust, 227
failure to use group resources, 227–228
misperception, 227
team leader and, 225–226
Diversity training, 232
Domain-relevant skills, 203
Dupont, creativity at, 211

E-mail:
communication norms and, 264–265
flaming and, 263
reduced social cues and, 264
virtual teams and, 270
Emotional distrust, 227
Emotional support, 69
Emotions:
e-mail messages and, 265
managing, 104
Empowered leadership, 169
Empowerment, 136–137, 242
Empowerment programs, 137–138
Enabling structure, 285
Equalization effect, 262
Equilibrium model of group development, 42
Ethnicity, as diversity type, 221
Ethnic minorities, in organizational
hierarchy, 220
Evaluation:
individual's, of benefits of participation, 43
motivation and, 60–61
See also Team performance evaluations
Evaluation process, creativity and, 204
Execution tasks, 26

Expectations, clear, 142
Expectations approach to diversity, 222
Extrinsic motivation, 204

Face-to-face interaction, 259
Factor jobs, 10
Factory teams, 30, 279
Favoritism, as problem in evaluations, 314
Feedback:
constructive, 103
from team performance evaluations, 310
Feelings:
managing, 104
See also Emotions
First impression error, 224
Flaming, 263
Force field analysis, 196, 199
Forming stage, 38–39
Free riders, 58
Fundamental attribution error, 224

Gain-sharing programs, 321–322
Gender:
as diversity type, 221
power styles and, 141
General Electric, core values and, 211
General Motors, Saturn
facility, 250, 279
Generation tasks, 26
Goals:
confusion with, competition
in teams and, 78
incompatible, 83
motivation and, 61
performance, 5
superordinate, 84
See Team goals
Goal setting, 296–297
GRIT approach, 119
Group(s):
characteristics of, 4
composition of, 23–25
defined, 4
teams vs., 5–6
traditional, 7
Group behavior:
types of, 67
See also Behaviors
Group cohesion, 61–64
building, 63–64
conflict resolution and, 63
cooperation and, 81–82
diversity and, 230
group performance and, 62–63

Team tasks:
empowerment levels for, 176
McGrath's model of, 25–28
See also Group tasks; Task(s)
Team training, 299–301.
See also Training
Teamwork:
benefits of, 242, 274–275
competencies and skills, 302
foundations of, 12–14
mixed-motive situation, 74
organizational culture and, 248–252
problems of, 275–276
See also Teams
Teamwork, stages of
alternative theories, 42
group development perspective, 38–40
implications of, 42–43
project development perspective, 40–42
Technology:
changes in, 11
job characteristics and, 10
Theory X, 9, 249
Theory Y, 9, 249
Third-party mediation, 120–121
Timekeeper, 66
Top management teams, 278, 282
Total Quality Management, 14
Traditional team, 7–8
Traditional work group, 7
Training:
assertiveness, 302–303
cross-training, 303–304
diversity, 232
interpositional, 303–304
team, 299–301
transfer of, 300–301
transnational teams, 253
types of, 301–306
Trait approach to diversity, 222
Trait approach to leadership, 169–171
Transactive memory, 300
Transactive memory systems, 50
Transfer of training, 300–301
Transnational teams, 252–254
miscommunication in, 253
Triplett, Norman, 14
Trust:
communication and, 98–99
cooperation, communication and, 85

Turnover:
diversity and, 228
of team members, 44

Unanimity, 129
Unhealthy agreement, 82
United States, organizational culture in, 247–248

Videoconferencing, 268–270
Virtual brainstorming, 206–207
Virtual teams, 257–272
advantages and disadvantages of, 265
brainstorming and, 266
change and development, 267
communication norms, 264–265
communication technologies for, 258–262, 268–270
conformity, 263
decision making and, 266–267
deindividuation and, 263
e-mail and, 270
miscommunication and, 263–264
organizational impacts of, 268
status differences in, 262–265
task performance in, 265–266
transnational, 253–254
Volvo, quality of worklife movement, 13
Voting, 151

Win-lose perspective, 119
Win-win perspective, 119
Withdrawal, assertive, 142
Women, in organizational hierarchy, 220
Work design, team performance evaluations and, 311
Workers, characteristics of, 9–10
Workplace, using teams in, 274–301
Work teams:
conflict in, 115–116
differences among, 277–282
evaluating effectiveness of, 276–277
problems with using, 283–286
production teams, 278
professional teams, 278–281
support for, 283
supporting, 283
top management teams, 278, 282
types of, 7
See also Team(s)

About the Author

Daniel Levi is a Professor in the Psychology and Child Development Department at Cal Poly, San Luis Obispo. He has an MA and a PhD in environmental psychology from the University of Arizona. He teaches classes in group dynamics and in social, environmental, and organizational psychology. In addition, he teaches classes in teamwork and the psychology of technological change in courses primarily for engineering students at Cal Poly. He has conducted research and worked as a consultant with factory and engineering teams for companies such as Nortel Networks, TRW, Hewlett-Packard, and Philips Electronics. In addition, he has worked on international team research projects in Europe and Asia.

Dr. Levi's research and consulting with factory teams primarily has focused on the use of teams to support technological change and the adoption of just-in-time and quality programs. This work examined a variety of team issues including job redesign, training, compensation, supervision, and change management approaches. His work with professional teams primarily has been done with engineering design teams. These projects examined the use of concurrent engineering, self-management, and the globalization of teams. The topics of this work included the impact of information technology on teams, facilitation and training needs for professional teams, and the impacts of organizational culture and leadership.

Early work on the present book was sponsored by an engineering education grant from NASA. This project focused on the development of teamwork skills in engineering students working on multidisciplinary projects. This project led to the development of cases and activities for learning teamwork skills and research on teamwork training and evaluating and rewarding student teams. Recent research on student teams examines cross-cultural issues and social support within student teams.

LaVergne, TN USA
12 May 2010
182369LV00003B/3/P